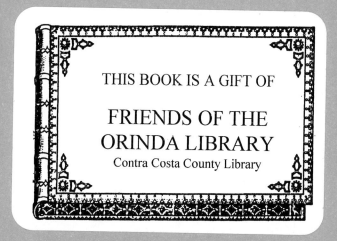

HANDMADE IN ITALY

HANDMADE IN ITALY

John Ferro Sims

text in association with Debra Boraston

Watson-Guptill Publications
New York

First published in the United States in
2003 by
Watson-Guptill Publications,
a division of VNU Business Media, Inc.,
770 Broadway, New York, N.Y. 10003
www.watsonguptill.com

First published in 2002 by
New Holland Publishers (UK) Ltd
Garfield House
86–88 Edgware Road
London W2 2EA
www.newhollandpublishers.com

Library of Congress Control Number:
2002109820

ISBN 0 8230 2189 0

Editor: Peter Kirkham
Senior Editor: Clare Hubbard
Designer: Debbie Mole
Production: Hazel Kirkman
Photographer: John Ferro Sims
Editorial Direction: Rosemary Wilkinson

Printed in Singapore

1 2 3 4 5 6 7 8/09 08 07 06 05 04 03

Page 2: Checco Lallo, Lazio, working on his pedal-
operated wheel.

Page 4: *top* Valter Solari, glass-based mosaic tessarae;
center Pino Valenti, *tarsia* and mosaic panel detail;
bottom Rafaele di Prinzio, wrought-iron head
boards.

Center: Ubaldo Grazia, ceramic tiles with hunting
scenes.

Page 5: *top* Antico Setificio Fiorentino, woven silk
ribbons and braids; *center* Barovier & Toso "mosaic-
effect" glass vases; *bottom* Artigianati Artesini,
painted goose eggs.

INTRODUCTION

This book celebrates contemporary artisans all over Italy who are not only employing traditional skills handed down through generations, but who are passionate about their work and dedicated to their labor. Whether working alone in a tiny workshop, with family members in a domestic studio, or with fellow workers in a small factory, these people are not simply making a living – they are creating a living out of a skill they have inherited or developed.

So, what constitutes "handmade" and how do you refer to the people who are featured in this book? They are definitely "makers": they use their hands to form, shape, press, weave, sew, carve, work with materials, and hold and use tools. Nearly all see themselves as artisans and, with characteristic modesty, deny that they are artists. Yet, when asked the source of their inspiration, they answer that it comes from their imagination, from within. It inevitably comes from the rich traditions of art and beauty so indigenous to Italian culture that it seems infused into their souls. So why don't they call themselves artists? Perhaps because they make a living from their skills and have turned their craft into a business. Or, more likely, because their products are mostly useful and practical items. After all, everybody needs plates on their tables, tools with which to work, and fabrics and carpets for their homes. No matter how fabulous their products –

be it luscious handprinted velvets, damask-steel knives, or filigree glassware – their origins are functional and practical. Handcrafted items give both the pleasure and beauty that art can offer, as well as the usefulness and efficiency that industrially designed goods provide.

Before industrialization, all tools and practical objects were made by hand and the craftsman was an integral, effective member of the community, central to the social and economic system. Nowadays, though, a handmade item is considered to be a luxury and, therefore, the prerogative of the more affluent sectors of society. Designs have evolved and have been perfected over centuries, and many articles that were made for absolute necessity are now desired and acquired mostly for adornment or embellishment – or even as a status symbol. A perfect example of such an evolution is the Sardinian shepherd knife: practical usage had dictated its form – the shape of the handle, the "give" of the blade, the curve of the steel. It has now reached a state of perfection both as a working tool and, in its most elaborate form, as a precious exhibit in the cabinets of private collectors the world over.

An artisan making something by hand creates an object that is individual. He or she conceives it and enjoys the satisfaction of seeing it through from beginning to end, taking pride in the final product. Craftspeople constantly develop their understanding of the material; when they begin a new work they regard

Top left

Coralli (Barovier & Toso, 1995). Designed by Daniela Puppa: a translucent-colored crystal incorporating fused shards of contrasting glass along the rim.

Top right

A leather bowl by Alessandro Marchetti with a stitched, decorated *bucchero* insert by Mara Santoni.

Bottom left

Fratelli Barato added colored enameling to copper objects by applying paint in powder form fired at a high temperature several times.

Bottom right

Detail of the aging effects resulting from the use of the *fresco strappo* technique by Roberta Ricchini of Open Art.

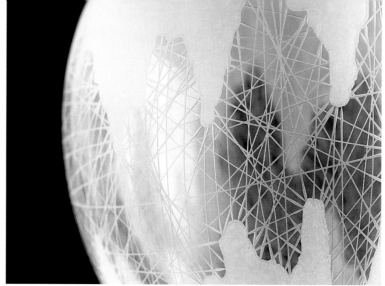

each plank of wood or slab of marble as a unique piece with its own character, smell, and organic make-up that has to be understood before it is modeled. Gabriele Buboli, a stonemason living and working in the hills behind Montagnana in the Veneto, contemplates huge lumps of stone with palpable delight before he even begins to tackle the task of carving it into a beautiful object, declaring, "There is everything in a lump of stone; it is up to me to find it."

Urbanization and technology are also eclipsing the craftsperson. Generally speaking, the younger generation does not seem to have either the patience or the staying power to embark on a lifetime apprenticeship to a chosen craft. In a button-clicking, e-commerce, global environment, where purchase is instant, variety apparently infinite (albeit dominated by brand marketing) and accessibility tantamount, there are few prepared to study a craft and perfect their skill over many years of repetitive labor with far from instant results. Many artisans savor the luxury of going slowly. Older generation craftspeople, who are eager to pass on their knowledge, bemoan the fact that the take-up of these skills in a revolt against mass production is still inadequate.

It might sound like a cliché to say these objects are labors of love — but that is precisely what they are. Why else does a Florentine silk weaver, who has perfected her craft over 35 years, stand hour after hour at an ancient loom strung with more than 48,000 delicate threads to produce just 8 inches (20cm) of fabric a day? Or a Tuscan sculptor stand knee-deep in alabaster dust carving a huge bird wing from this rare raw material?

Invariably, they have all chosen or even "found" their occupation and none of them have deviated from it; some argue that their skill is genetic. Sons of artisan fathers who have not been pressured into the family business have, nevertheless, instinctively developed an interest and found they have a natural skill: examples include the maverick terracotta artist Luca Vanni in Impruneta who gave up biking, truck-driving, and other wayward occupations to follow in his father's footsteps. The son of woodcarver Mastrenrico in the Veneto region of Montagnana is steeped in his family's business of basket weaving and woodcarving and can envisage no other career for himself but to carry his inherited skill through to the next generation. Both are touchingly proud of their fathers' craftsmanship. Similarly, daughters of weavers grow up mastering the skill of the cloth without questioning their choice of career.

Others, inspired by their own desire to work with their hands, or by the traditions around them, have apprenticed themselves to masters of their art: leather designer Patrizia Pinna worked with an elderly artisan in her local town making standard leather handbags and wallets for 18 years before branching out on her own to create the fantastic bags, book covers, and other goods she now makes in the picturesque Tuscan hill town of Sorano; and in Massa Marittima Alessandro Marchetti, after a similar apprenticeship, now combines his skills in leather design with those of Mara Santoni who has adopted the ancient Etruscan method of ceramic – *bucchero* – to make highly original plates, bowls, vases, and other decorative objects (he tells an extraordinary story of a 90-year-old woman rushing into his workshop, having heard what he was doing, quite literally with the gift of her knowledge and skills in this traditional craft); or Valter Solari working in the countryside near Spilimbergo, Friuli, where the tradition of mosaic work is intrinsic to the area's artistic wealth, who was naturally drawn to the craft and could not envisage any other occupation. The Venetian glass-making island of Murano is probably the best example of a situation in which generations and generations are still steeped in a craft tradition that goes back nearly a thousand years and where even those who move to other cities take their artistry and skills with them.

These arguments can be applied to craftspeople the world over, but they are enhanced in Italy by the rich cultural heritage and the history of art. You only have to wander the streets of Florence or Venice to appreciate the wealth of small workshops where the noise and the products of hundreds of craft shops spill through the doors into the streets – as they have been doing for centuries. The Renaissance was the boom time for the arts in Europe, and Italy is acknowledged as the supremacist in the major arts of painting, sculpture, and architecture. Evidence of this still abounds, not just in the major cities, but in small historic towns all over the country, and has infused its inhabitants with an almost laissez-faire acceptance of their glorious past. In America and the majority of northern European countries, art is prized and preserved behind glass or in museums, but the Italians live with their culture in everyday life – it is all around them.

However, this monumental display of artistic heritage is only part of the story. At the same time as the great masters, the artisans were also flourishing. The goldsmiths and silversmiths, the silk weavers, potters, jewelers, glassmakers, and furniture makers were also toiling to create a resplendent way of life for society at every level, including the practical and functional. Artisan crafts reached new heights of excellence during the Renaissance, and the styles and designs are still preserved and produced today.

Artists and artisans worked together. Not only were the craftspeople turning out works of art, but celebrated artists were also working in the world of the applied arts and creating functional and practical objects and designs. How many people are familiar with, for example, Donatello's designs for stained glass, Michelangelo's designs for the reading desks in the Laurentian Library or his inkwell for the Duke of Urbino, da Vinci's design for a warping machine for the Antico Setificio in Florence, Titian and Tintoretto's designs for mosaics and tapestries, or Francesco Guardi's leather wallcoverings in the Redentore in Venice?

Italians are supreme individualists and are continuously reinventing themselves. Corporate culture remains peripheral. On the whole, the Italians have spurned the chain stores and fast-food outlets that proliferate in the rest of the western world. Every restaurant and coffee bar and the majority of fashion boutiques in every Italian city remain determinedly individual – mainly independently and family owned. Even the large fashion houses still have a central, controlling figure at the helm making sure their product has a hallmark of

9

individuality, be it the Alessi, Ferragamo, or Versace family or Giorgio Armani himself. And when they have established success, Italians do not rest on their laurels and keep churning out the same formula. Experimentation and reinvention drive their creativity forward. In the same way, today's artisans are using the traditional skills they have inherited not only to preserve and restore old styles, but to innovate and create fresh expression.

Such individuality has its roots in political, social, and economic history. Italy's status as a unified country is still relatively recent (less than 150 years), and people today still refer to themselves as natives of their city or region. Florentine, Venetian, Roman, Sicilian – these identities are much more definitive than merely "Italian." Historically, Italy's republics and city states vied with each other for supremacy. Religion and politics all played a hand, but one of the most important areas of competition was trade. From the 12th century onward, artisans organized themselves into guilds with the dual aims of protecting their inherited knowledge, which was jealously passed down from generation to generation, and of continually improving the high quality of goods produced. The Venetians were at the forefront, and even today the younger generation artisans in Venice, who are outside the family business, are aware of a reluctance of the older masters to pass on the secrets of their trade. Indeed, Venice is the best, if the most extreme, example of this theory.

The artisan guilds, or *scuole*, were the backbone of Venetian society. When Napoleon arrived at the end of the 18th century, there were more than 300 of these corporations in existence – from the glassmakers' and arsenal workers' large and politically influential guilds to the humble fruit dealers' guild. At the height of La Serenissima's power, the artisans flourished and preservation of their power was tantamount. Anybody who left the city to work abroad was, at best, refused re-entry, at worst tracked down and murdered. Only the most revered in their trade could hope for special dispensation, one example being three highly regarded mirror artists from Murano who were taken to Versailles by Louis IV, but had to obtain a special pardon to be allowed to return home.

The positive aspect of a country with so many competing identities is that societies developed individually and independently. This development was compounded by the geography of the country, which made transportation difficult. Over the centuries, there also grew a marked difference between the north, with its artisan middle classes and greater democratic power, and the south, which remained under the control of feudal overlords or absentee kings.

In the 19th century energies, were being directed against external domination and toward political unification and, unlike in Britain where the industrial revolution was in full swing, the Italian economy desperately lagged behind. The two world wars high-

10

Previous page (left)

Weaving is labor intensive work, especially with fine textiles, such as this velvet damask at Tessitura Bevilacqua in Venice.

Previous page (right)

A modern engraved glass bowl from the Varisco studio in Treviso.

Opposite (top)

Marcello Patrì from Caltagirone, Sicily, has evolved an antique style of painting and distressing.

Right

Hand-woven silks in the great Italian Renaissance tradition are still being produced on antique wooden looms at Antico Setificio in Florence.

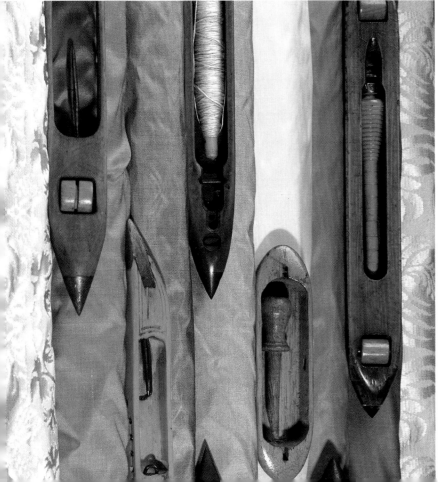

lighted Italian industrial weakness and pressure to catch up with northern European and American standards did not accelerate until the 1950s. From the "handmade" point of view, the late arrival of industrialization can be seen as a great blessing. Even as the machinery moved in after World War II, there were still huge numbers of craftspeople providing handmade goods, and, consequently, there are still older generation artisans alive and available to teach the younger generations.

In many advanced western countries, the modern day artisan has become something of a protected cultural reserve. Whereas in the Renaissance, guilds were set up to promote and maximize the skills of their members, associations are nowadays introduced to save and preserve what is seen to be a dying breed – even in Italy – and the products of these artisans are often promoted along with tourism to encourage visitors to go to the workshops alongside their other sightseeing.

11

I have sought out individuals and small concerns all over mainland Italy and the islands of Sicily and Sardinia, visiting artisans in their workshops, who are dedicated to creating objects of beauty, quality, and utility, while preserving tradition and techniques handed down through the generations.

This book cannot show the whole spectrum of "handmade in Italy" without becoming an encyclopedia, so the criterion of choice have been limited mostly to items that have their origins in the home, rather than personally decorative items such as jewelry and clothing. The selection is also necessarily subjective. The book must, therefore, be seen as a sample of major, and a few minor, artisan businesses across as many regions as possible. I hope that you will be stimulated by this book to explore firsthand Italy's "handmade" heritage.

stone ~ from tomb to table

Italy is geologically blessed with a wide variety of stone: sedimentary limestone, in all colors, from the white of Martina Franca and soft brown of Lecce in the south, to the *rosso* (red) of Trento and the creamy white of Montagnana in the north. Sardinia is the main source of granites, and central Italy provides the volcanic rocks, the peperino, tufa, and travertine, as well as the softer alabaster and famous white Carrara marble, the purest form of metamorphic calcium carbonate (limestone). Working with stone, anywhere in Italy, still means essentially working with local materials. The Etruscans were the first great technicians in the art of working stone; the pragmatic early Romans concentrated on construction. They developed a building system based on stone and brick that remained in general use in Italy until the beginning of the 20th century.

Arguably the greatest artist produced by Western civilization, Michelangelo brought to sculpture the complete mastery of the human form in stone. Decorative stonework is, of course, not just used to embellish façades, to create reliefs or, indeed, to form sculpture; the tradition of mosaic floor pavements reached Rome from Greece.

Of all the traditional artisan skills covered in this book, it is the stone-mason, the mosaicist, and the sculptor who are facing the greatest problems today. The recent introduction of reinforced concrete dealt a severe blow to all traditional builders and stonemasons, and the hard-acquired skills of these workers' became virtually redundant.

opposite: Cavaliere Ferdinando Palla. Three stemmed classical bowls illustrating the use of marbles with unusual coloring. Two are made of French red marble and the third (foreground right) is of *Breccia di Siena*. The background is an elaborate *pietra dura* table.

MATTIVI MARMI, STONEWORK, TRENTINO ALTO-ADIGE

"I am particularly pleased that local communities are once again commissioning fountains for la bellezza del comune *— both as symbols of beauty and to remind the young of a heritage not forgotten by their grandmothers."*

ALESSANDRO MATTIVI

15

THE STORY OF THE MATTIVI FAMILY in Trento illustrates the vicissitudes of the stone business. In 1875 the family bought some land on the far outskirts of Trento by the mountainside that had been quarried for centuries. *Il rosso*, the red sedimentary rock known to the Romans, was a particularly good facing material and could also be sculpted quite easily. As elsewhere in Italy, the early years were good – the boom in demand for new buildings had arrived, following the optimism that the unification of the country had brought.

Both the original founders of the Mattivi business had been *scalpellini* (stone carvers) – hand cutting slabs from the rock face, then carving them into the desired shape and form for their clients. The sudden cement-driven crisis in 1914–15 meant that those who wanted to survive in the business had to adapt. They turned to sculpting as a means of survival, contenting themselves with funerary work – Madonnas, Christ figures, and lifelike images of the dead.

For the few survivors in the old traditions, times are improving. Despite the fact that the Trento city boundary has now reached their workshop under the crag of their old quarry, the Mattivi family is very active in all areas of domestic and architectural design, making one of a kind items. They stand out in a sector that produces factory-made series. Alessandro, the grandson of the founders, has even returned to his roots in one particular sense. He no longer buys blocks of stone from stone merchants, preferring to choose blocks from the quarry himself to avoid the problems of stone defects. After all, it is not easy returning 5 or 10 tons of blocks from whence they came! This way he takes the stone through all of its various stages to the finished product and has greater flexibility in changing elements in the design along the way. Alessandro is encouraged that local communities are again commissioning fountains in *rosso* for public places.

Above

A neo-Romanesque capital carved by Enrico Mattivi, Alessandro's father, in the days when a *scalpellino's* work was appreciated.

Opposite

One of the fountains recently commissioned in the local *rosso* stone of Trento by the small nearby community of Ravina.

Below

One of Giovanni Martellini's interpretations of the fabled mythic Lion of the South. This alert, amusing, and thirsty limestone creature with a particularly comic moustache has not yet found a home. Giovanni is happiest when he allows his whims to take over the chisel.

16

Above

Giovanni Martellini's simple, elegant representation of the sun carved in the local, virtually white, Murge limestone.

Opposite

A corner of Giovanni Martellini's workshop at the limestone quarry near Martina Franca, Puglia, where various slabs of stone are stacked in an apparent abandon.

GIOVANNI MARTELLINI, STONEWORK, PUGLIA

"Around me in Martina Franca I find inspiration: the palaces, churches, and portals all provide me with great examples and challenges."

GIOVANNI MARTELLINI

AT THE VERY OPPOSITE end of Italy and on a very different scale, Giovanni Martellini is a true *scalpellino* of the old type. Even his surname, the plural of small hammer, fits him perfectly. Giovanni works in a large quarry near his hometown of Martina Franca, in the Murge region of Puglia, producing pediments, gargoyles, lion heads, and architectural items for local builders in his little corner workshop. He works alone in the midst of the dust and chaos as huge trucks go back and forth with the quarried limestone.

Stonework is a long-standing tradition in the upland limestone plateau of Puglia. A quarry can be opened practically anywhere in the region, and stone is ever present in town and countryside alike; from the strange beehive buildings called *trulli*, to the dry-stone walls containing small plots of land, to the many beautiful and historic buildings of the Murge. There is no shortage of workers in stone.

Giovanni is intensely passionate about stone and about his birthplace. He talks of the smell, the feel, the textures, the colors and of what can be found in the stone – what has grown and maybe even died therein. He speaks softly, gently, and describes how, at age 11, he started to carve in stone.

From his first boyhood passion, it took Giovanni more than 10 years to learn enough about stone and how to work it before he could properly call himself a *scalpellino*. For the next 23 years, he made a living carving headstones for a local cemetery, learning his craft, and developing his own ideas. He now makes a living both from his day work at the quarry and by working in his own little workshop, near his apartment, in the historic town center of Martina Franca. Giovanni says: "An artist would die in a hole in the wall if he tried to survive only on what he wants to make."

In his workshop he has a collection of carved stone body parts – feet, hands, arms, lips, heads – and some beautifully carved table feet of differing designs. He also makes or modifies most of the special tools he requires. Even his home is a showcase for his work with lamps in white stone, elegant fireplaces, and small, simple stone birds adorning the walls.

LABORATORIO DEL MARMO, GABRIELE BUBOLI, STONEWORK, VENETO

Above

Gabriele Buboli's workshop is at the back of his garden and the path from the house passes vines with his interpretation of the sun carved on a rock found nearby.

"There is everything in a lump of stone – you just have to find it."

GABRIELE BUBOLI

BASED NEAR THE MEDIEVAL walled town of Montagnana in the Veneto, Gabriele Buboli has much in common with Giovanni Martellini. Gabriele is another passionate *scalpellino* who has been involved in all areas of stonemasonry for the past 40 years. He, too, made headstones and funerary portraits for a living and spent more than 10 years in Switzerland following his apprenticeship in the quarries near his hometown. In his spare time he sculpted and, actually, is a very fine, but modest, sculptor.

Today, Gabriele and his son work in a small studio in the garden behind their house, mainly creating table bases in local Vicenza stone (a cream white limestone) for a nearby company called Laboratorio del Marmo. This company has a unique way of mixing the artisan and the marketplace that is commercially viable. Founded in 1990 by Giovanni Andrian, a furniture maker, and a local sculptor Giandomenico Sandri, they decided to make furniture items that were individual and handmade by establishing a network of local artisans dotted around the Montagnana area to carry out the work.

All of the artisans, including Gabriele Buboli, start with a general design that is roughly sketched out on the surface of the stone block. The process of carving is begun using hand power tools; then the sculpture is worked on with hammer and chisel. As with all items essentially made by hand, the company is constantly evolving new designs combining organic shapes for the stone bases with original glass shapes on top. Glass is the preferred material because it enables the design of the base to be seen, but wood is also used – again in an unusual and interesting way by integrating it with stone. The shape of the base is actually cut out of the wood to reveal something of the character of what lies beneath the table's surface. Indeed, Laboratorio's sophisticated catalogue is seen as a starting point for discussing new and unique designs.

Left

The form to be sculpted starts with a design sketched onto the stone block from a clay model. Gabriele Buboli continuously refers to the model as he carves with a hammer and chisel to create the detail. The finished object is lightly rasped to smooth out the rougher edges.

Above

A selection of limestone bases to support glass or wooden tabletops: made by Gabriele Buboli and his son for a nearby company, called Laboratorio del Marmo.

Above

A classical bust of a siren with serpents by a local sculptor. The white Carrara marble is often referred to as *marmo statuario* (statuary marble).

CAVALIERE FERDINANDO PALLA, MARBLE AND PIETRA DURA, TUSCANY

"The days of the great sculptor in stone are probably over. Few youngsters are now willing to devote upwards of 20 years to learn this wonderful skill."

AMELIA PALLA

PIETRASANTA, IN TUSCANY, sits at the foot of the Apuane Alps, for centuries the main source of the famous white Carrara marble, first used extensively by the Romans. Michelangelo Buonarroti, whose ability to extract an animate form from a block of stone was wondrous, was above all a *scalpellino*. A carver in marble, whose abilities remain unsurpassed, he chose Carrara marble for many of his greatest works, including the stunning *Pietà* now in Saint Peter's Basilica, Rome, the contract for which was actually signed in Pietrasanta in 1498.

In the many towns on the edge of the Apuane Alps, the long slow decline in marble use, which began at the end of the Renaissance, came to an end with the unification of Italy. In the small town of Pietrasanta, one of the oldest company names is that of Cavaliere Ferdinando Palla, who was working as a stone carver before the end of the 19th century. In the early years, Ferdinando and then his sons, Spartaco and Ugo, followed the tradition of copying classical and neoclassical statues, both for private clients and for the church (then a good source of work). After World War II, a shift occurred toward architectural commissions from local communities and clients. Before the war, a sculpture studio might have employed a series of specialist workers devoted to each individual phase in the making of a sculpture: the "artist" chose the block of stone; then one worker would create the outline, while another would bring the figure into being; then the "artist" would bring the stone to life by creating the face, hands, and flesh tones. Another specialist formed the clothes, fabrics, and other objects and yet another would create any animals, birds, or flowers that were required in the composition. Finally, the whole work was cleaned and made as

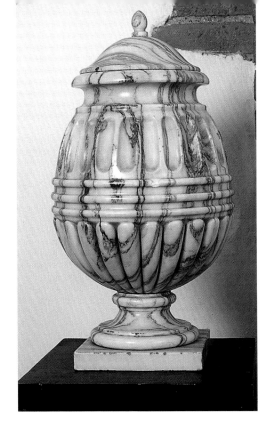

Left

A classical urn in a local Apuane marble known as *fantastico*, literally, "fantastic."

Below

A marble table edged with mosaic of unpolished marble tesserae.

perfect and smooth as possible by a worker whose sole job was to prepare the finished sculpture for the client.

Those days are over and virtually all sculptors are now responsible for their output from start to finish. In the old tradition, though, is the creation of a full-scale three-dimensional reference model in gesso or plaster from which the sculptor then works.

Now into the fourth generation, Amelia and her brother Antonio, carry on as best they can in the traditional Palla way, even though the family has split up into various rival companies. In their somewhat chaotic workshop, employees and visiting artists work on different projects, from small sculptures to architectural columns and pediments, to restoring inlaid hardstone tables. As well as satisfying domestic and architectural clients, they are still sculpting fine marble objects that testify to their innate abilities and artistic imagination.

21

ALAB'ARTE, GIORGIO FINAZZO, AND ROBERTO CHITI, ALABASTER, TUSCANY

STILL IN TUSCANY, Volterra, some 48 miles (80 km) south of Pietrasanta, is a city whose fame is based on two factors: it was one of the major towns of the Etruscan League and, uniquely in Italy, it sits near mines of natural alabaster, a dense form of the mineral, gypsum. The name "alabaster" is thought to derive from the Egyptian town of Alabastron, perhaps the source of a similar material called onyx marble, used to make vases and amphorae. Alabaster is much softer than marble, is generally translucent and white, cream or gray, and can be easily carved with a knife and polished to give a smooth, waxy appearance.

"We strive to maintain the very best traditions of alabaster work to ensure that this marvelous material is properly appreciated."

GIORGIO FINAZZO AND
ROBERTO CHITI, VOLTERRA

Giorgio Finazzo and Roberto Chiti were born and raised in Volterra and are genuine Tuscan craftsmen who joined forces to create fine items in alabaster. Although it is not difficult to carve, like all stones it needs to be understood to be worked well. Together, as Alab'Arte, they have helped to raise standards; apart from being great copyists, they are regularly commissioned to sculpt decorative objects such as lamps and vases, as well as original work of their own design.

Opposite

Two elegant, sensuous, refined examples of exquisite alabaster objects from Alab' Arte. A gallery in San Gimignano commissions their more original work.

Left

Alabaster globes of various dimensions underlit to demonstrate the translucency of the material. Volterra is the only place in Europe that has natural alabaster. The almost transparent kind is called *Bardiglio*.

23

CLAUDIO BERTI, WORKS IN PIETRA DURA, TUSCANY

PERHAPS THE MOST extraordinary and completely mystifying of all of the stone-based "arts" is that of *pietra dura* (literally "hard stone," meaning a type of mosaic executed in semi-precious stones). Unbelievable levels of skill were achieved in Florence under the patronage of the Renaissance Medici family. In the 17th century, designs mostly featured flowers and birds but later moved on to landscapes and scenic views. The Montelatici family in Florence, active in the late 19th century and early 20th century, produced extraordinary works in *pietra dura* that imitated genre paintings of the period.

The inheritors of the traditions of the Medici Grand Ducal Workshops in Florence are now small companies such as the Berti family who have a tiny shop front in the town and a workshop on an industrial estate at Scandicci, a nearby suburb. The father, Giuseppe, was 20 years old when he started the business just after World War II. For seven years he cycled 5 miles (8 km) to

"It's not just about finding; it's about knowing."

CLAUDIO BERTI

catch the train into Florence, worked for 12 hours, then repeated the journey home. At first he worked with his uncle, then a cousin, and when he died, Giuseppe took over the business with his brother. His son Claudio now looks after the five workers, and the company receives commissions for new work as well as restoration projects from all over Italy and from international clients in London and New York.

Passion for the work drove Giuseppe forward. He still regularly spends time wandering around the Tuscan countryside digging for stones, looking for that special something that a composition needs. He appreciates the textures, the variety of colors and degree of polish that they will take. One particular *pièce de résistance* for Giuseppe was a 17th-century table design that he copied. This required several panels featuring birds and flowers, with three key round "pictures" featuring monuments. He adapted this for an Arab customer, featuring the Sheik's own palace

24

Left

A detail of one of Berti's *pietra dura* tables.
A large number of different stones are utilized in
the outer ring. The hundreds of stones they use
come from all over the world: lapis lazuli is the
most expensive, the more dense the blue, the
more precious the stone.

Below

Giuseppe Berti working on one of his stone
paintings based on Pissarro's *La Route de Gisors à
Pontoise*. By the time it is finished, it will have
taken nearly six months' work.

Opposite

This small cameo *pietra dura* featuring roses and
lily of the valley took Giuseppe Berti about 10
days to make as a little gift for his wife.

in the center panel and two castles in the two smaller flanking panels. From Giuseppe's
perspective, though, the future looks bleak: "It takes years in the workshop to acquire
the necessary skills for this work and the young are no longer willing to make these
kind of sacrifices. This type of work will soon die unless something is done to reverse
the situation."

Even though Giuseppe will take his inspiration from anything – books, photographs, copies of masterpieces – he definitely has his own style and, after being there
a while and studying some of the "stone paintings," it becomes clear which ones are
his. As he is "officially" retired, he often spends time making gifts for his wife – the
small cameo brooch (pictured opposite) for example, took him about 10 days to make.

Stones used in their designs come from all over Italy and from elsewhere; lapis
lazuli from France and Persia; chalcedonies and alabaster from Volterra; various granites
from Sardinia; red marbles from the province of Siena; black marble from Verona;
different jaspers from Sicily; marble granites, gray malachite, and other minerals from
Switzerland, are just some of the hundreds of varieties that they keep in stock.

Top

Valter preparing the ground for a large
wall-mounted mosaic.

Above

Valter working with dark tesserae for a small mosaic.

Opposite

A detail of tesserae used in modern mosaic
techniques, where volume is an essential part of
the visual appeal.

26

VALTER SOLARI, MOSAICS, FRIULI VENEZIA-GIULIA

*"Nearly all of my work is commissioned, 70 percent of it has religious subject matter,
especially Christs and Madonnas. When I can, though, I prefer to do portraits of friends
and family, where I can use more expressive elements than in the more commercial work."*

VALTER SOLARI

MOSAIC IS THE ART of embedding small pieces of cut stone or pigmented glass in
a plaster bed to serve as floor or wall decoration. Solidity, resistance to moisture,
durability and color-fastness make mosaic a practical form of architectural decoration
in the warm and humid areas of the Mediterranean. A mosaic begins with small squares
of cut stone, pigmented glass, or gold or silver leaf sandwiched by glass, known as
tesserae. In preparing a ground for the tesserae, the mosaicist usually coats the surface
with three thin layers of mortar; on the second layer, the major lines of the design are
sketched as a guide for embedding the tesserae in the damp plaster of the third layer.
The sizes, shapes, varieties, and color intensities of the tesserae, as well as their patterns
of insertion, vary according to period, place, and required effect.

The process of making opaque glass tesserae was developed 2000 years ago by the
Romans by fusing silica and various minerals and then pressing the resulting
compound between marble slabs to form a "pancake" or "pizza." Apart from a higher
fusion temperature to produce a more uniform product, the modern method of
production has remained the same.

Unknown to most Italians, and certainly to the rest of the world, is the *Scuola
Mosaicisti del Friuli*, founded in 1932 in Spilimbergo, in the northern province of
Pordenone, Friuli Venezia-Giulia. During his foundation art course there, Valter Solari
realized that the craft of mosaic suited him perfectly, so he embarked on the three-year
course and graduated in 1981. After a period in Paris, military service, and seven years

28

at one of the greatest Milan-based mosaic laboratories, he decided to return home and set up a workshop in Dignano in 1990 with his wife, now also a teacher at the *Scuola Mosaicisti*. He furthered his knowledge of ancient techniques and restoration with experts from Ravenna at sites in Aquileia and at El Jem in Tunisia.

Valter is one of life's pragmatists in that nearly all of his work is done on commission, with 70 percent on religious themes for local churches. He is able to execute faithful copies of details or sections of floor mosaics of the Hellenistic or Roman periods and copies of Byzantine icons as well as interpretations of Renaissance and contemporary art. However, he is keen to point out that this is not necessarily what he would choose to do, but he has to make a living and that this is the way of the market. So, when pushed, Valter admits that he enjoys working on other things and soon enthuses about the 12 works based on the paintings of Klimt that he is preparing for a huge mosaic art show in Burgundy. These works show the higher relief and three-dimensional possibilities of mosaic as well as the range of tonalities produced by contrasting stone and glass-based tesserae. The fine portrait of Valter's wife's grandfather, carried out in the more modern style further demonstrates Valter's ability to give shading, form, and character by varying the size, position, and choice of the tesserae. Viewed from a little distance, the eye fails to perceive that the portrait is in stone at all.

29

Opposite

A mosaic in the flatter Roman-style where opaque tesserae are placed much closer together to create a two-dimensional mosaic.

Above

A fine portrait of Valter's wife's grandfather, carried out in the more modern style.

ARTE BARSANTI, PLASTER OF PARIS SACRED STATUES, TUSCANY

GYPSUM IS THE MAIN ingredient of plaster of Paris, and plaster-cast statues cost much less than their stone-carved equivalents. Around 1900, Carmelo Barsanti founded a small industry specializing in sacred statues in plaster in the small town of Bagni di Lucca, just north of the Tuscan town of Lucca and not far from the famous marble towns of Pietrasanta, Massa, and Carrara. It appears that the fashion to have busts and portraits in *gesso* (plaster) emerged in Italy after the Napoleonic period, and probably because of the huge demand generated by the optimism of the unification of Italy, many workshops opened near the major "stone" centers to pick up the cheaper work for which sculptors could not compete.

At first Carmelo produced small statues in varying styles but dropped secular work after the end of World War II to concentrate on sacred statues. All the individual figures were, and still are, sculpted by local artists. From these originals, molds are made which are then used to make the statue copies: up to 1,000 copies per mold is possible, with care, with the first always kept as the *campione* (master sample). Originally, specialist painters were employed; some for decorating the skin tones and others who painted the clothes. The most elevated were those who painted the eyes – especially of the larger statues. Today, glass eyes are used on all statues from 32 to 52 inches (80 to 130 cm), the maximum

"I am trying to maintain the traditions that were initiated in Bagni 100 years ago by my great-grandfather Carmelo Barsanti. We are the only ones left who make sacred statues in plaster."

SIMONE FIORE

size made. Painting was traditionally done with powdered colors mixed with linseed oil and applied directly to the fresh plaster but nowadays the plaster is first made waterproof, then decorated with standard colors. A large statue can take up to one and a half days to paint.

As a result of the introduction of plastic, the 1960s saw a great decline in the market for plaster statues, and all the small family-based workshops around Bagni suffered. Plastic statues are much cheaper to produce, especially in quantity. This has left Barsanti as the only workshop still making statues in plaster.

Simone Fiore, Carmelo's great-grandson, took over the running of the workshop in 1996 and is trying to maintain the traditions initiated in Bagni 100 years ago. There are some concessions to modernity – such as making molds out of resin-based materials so that copies remain closer to the original for longer. But essentially the process remains the same; plaster is mixed to the right consistency, then poured into the mold, and allowed to partially dry for about 20 minutes, so the excess can be raked off from the statue base. The casts are then transferred to a slow oven, really just a warm room, where the plaster is allowed to dry out completely for about a day. They are then cleaned of all surface imperfections, ready for their final decoration.

Left

The Barsanti
workshop in Bagni di
Lucca is virtually a
museum. Hundreds of
statues, molds and
casts are littered
around the building,
filling every available
space. The extent of
this Barsanti legacy is
not known because
no inventory has ever
been made.

31

ceramics ~ an art of necessity

Three distinctive types of wares are now confusingly referred to by the generic words "pottery" and "ceramics." Earthenware, or terracotta, is composed of clay, or blended clays, which is shaped into a form then hard baked, usually in a single firing to make the *cotto* (a shorter version of the Italian word *biscotto* or "biscuit," the brownish red color of baked earth). Stoneware is made of clay containing high levels of alumina, which is fired to much higher temperatures and is usually glazed. The third type is porcelain, invented by the Chinese, using a similar material, confusingly called "China clay" in the West.

Traditionally, Italians only make pottery from ordinary clays which, with the addition of tin-oxide glazes (introduced sometime in the late 14th century), came to be known as *maiolica* (also spelled majolica in English). From Faenza, the Italian city historically celebrated for the production of *maiolica*, comes the term *faience* which refers to a specific type of opaque tin-glazed earthenware that became popular throughout Europe from the 16th century, but as the century came to an end, porcelains especially from France and England began to dominate the market for tableware.

opposite: Mastro Cencio's "high-Renaissance" plate with a representation of an *gattopardo* (leopard) often depicted in idealized hunting scenes, echoing the style of the late 1400s, early 1500s, Deruta, Umbria.

ALESSANDRO LAI, ARCHAIC POTTERY, SARDINIA

"My archaeological studies have inspired me to develop my own designs which echo the past faithfully."

ALESSANDRO LAI

Above (left)

A flat-based bowl with a simple graphic motif that reinterprets Arabic influences in early Sardinian pottery.

Above (right)

Some of the earliest ceramics were decorated by pressing shells into the damp clay before firing.

HISTORICALLY, THE CERAMICS of Sardinia have been practical and functional: simply decorated terracotta containers for oil, water, and wine. Alessandro Lai, a young potter from Iglesias, became particularly passionate about local history and the evolution of pottery in his area after studying at art school in the capital, Cagliari. Over the years he has taught himself enough about ancient techniques to be able to make for his own collection, and for various Sardinian museums, facsimiles of ancient pots, bowls, and vases.

He has also reinterpreted ancient designs as well as reintroduced *bucchero* ware, the process whereby clay is fired in a wood kiln in such a way as to allow the smoke to turn the clay black. Unusually, he adds white designs to the *bucchero* before a second firing. It is hardly surprising that many potters in Italy, like Alessandro, start by copying the past, when the past is so rich and varied and when, in reality, every type of design and form has already been created and perfected, not once but maybe dozens of times. Although he remains essentially an artisan, someone who has to make useful objects for a living and who rarely strays into what is usually termed "art," Alessandro is evolving his own styles in parallel with his more "academic" creations.

Left

One of Massimo Boi's modern "plates" achieved by a combination of techniques, two firings and special paint colors.

MASSIMO BOI, MODERN CERAMICS, SARDINIA

"I like to use very primitive 'tools' to achieve sophisticated results. Experimentation and expression are my masters."

MASSIMO BOI

35

WHILE ALESSANDRO LAI is re-creating the ancient past, not so far away is another potter with a contemporary edge, Massimo Boi, who cares little for the practical and has developed his own style. Massimo studied the history of art at Oristano before returning to Quartu Sant'Elena, near Cagliari. His unique designs are inspired by the environment. He uses flowers, animals, and the sea as well as architectonic elements to tell a story. He also uses special clays not found locally; they come from Vicenza and can withstand the rather primitive and improvisatory techniques he employs. After the first firing, his plates, pots, and vases are decorated with colored glazes that are then subjected to trial by fire – by partially burying them in straw and setting light to it. The unglazed clay turns black, the colors run and fuse and, once an effect that pleases him occurs, he douses the pot with cold water to halt the process. Ordinary clays would crack instantly at the sudden change in temperature, hence the need for special material. Massimo's technique imitates raku ware, but the glazes he uses do not craze to give the characteristic crackle appearance. Massimo's eclectic style and hunger for exploration has resulted in a wide range of highly charged, dramatic designs.

STUDIO D'ARTE MASTRO CENCIO, CERAMICS, LAZIO

"I follow my passion for a style until that avenue of evolution is exhausted, and then I move on to the next one that comes my way which interests me."

MASTRO CENCIO

BACK ON THE MAINLAND, north of Rome in the town of Civita Castellana, is a potter who has learned to master every technique discovered in Italy since the dawn of time. Vincenzo Dobboloni, known as Mastro Cencio – *Mastro* because he is a true master of the potter's art and *Cencio* because it is a diminutive of Vincenzo. His father was a pastry-maker who, like many others in this very Etruscan part of Italy, was an accidental *tombarolo* or tomb robber. Often, when out picking wild asparagus or mushrooms in the deep ravines prevalent in this area of the Tiber valley, he would find and collect ceramic pieces. When Vincenzo was about 12, his father gave him a small statue of a female figure he had unearthed. For some reason,

Left

One of the copies of an early Hellenistic *figura nera* (black figure) two-handled vase that is virtually indistinguishable from the original. Mastro Cencio usually includes small "incorrect" pictorial elements to prevent him from being accused of "fakery".

Vincenzo was totally smitten with the statue and carried it with him everywhere. Vincenzo's father already appreciated his son's drawing ability and, seeing that the boy was so keen on his find, suggested he should study pottery. Vincenzo jumped at the opportunity and went to the workshop of the most important local ceramicist – Dino Dominicis, who knew everything there was to know about ceramics. Vincenzo initially threw himself into the Etruscan style, from which he developed his interest in making copies of old pieces.

Although it is generally accepted that a modern ceramicist needs to understand chemistry in order to evolve, experience, dedication, and a passion for research over the past 30 years has enabled Vincenzo to mimic every ceramic style from the past 3,000 years. With the appropriate colors, clays, and decorating techniques, he can make virtually perfect copies of any pot, so much so that only carbon dating shows his works to be new. However, his motivation has never been to cheat or defraud his clients. Vincenzo seeks instead to enter into the mind and spirit of the original artisan so as to understand how the object was conceived and then created. Colors especially present him with challenges but these challenges are relished and, in the end, are overcome by his diligent research and experimentation.

Above

A beautiful Hellenistic "kilix" minotaur pedestal bowl in the *figura nera* (black figure) style with scenes of hunting and eating. It takes remarkable skill to make and decorate such an object.

CHECCO LALLO, RUSTIC POTTERY, LAZIO

"I am the last of the pignattari *in Vetralla and when I die a tradition will come to an end."*

CHECCO LALLO

IN NEARBY VETRALLA, an old Etruscan town west of Civita Castellana, lives an older potter called Checco Lallo who is the antithesis of Mastro Cencio. Fifty years ago there were about seven potters still working the Via dei Pilari, a country road that circles the southern side of the town. In the past, Vetralla's potters started work at about eight years of age within the family-based concerns, and the different families vied with one another for business. Young boys had the hard job of preparing the clay that the men had excavated and carried back from nearby Monte Panese for throwing; the women were responsible for decorating the pots and applying the glazes. They produced black glazed ware, by adding manganese to the glaze, and the more traditional red. These potters were known as *pignattari* from *pignatto*, the terracotta pot once used to cook beans.

The name by which Felice Ricci, Vetralla's last potter, goes by is that of his grandfather, Checco Lallo. All the family potters were known by this name, which was passed down from father to son. Now in his seventies, he doesn't speak when he works, preferring to concentrate on the motions that he knows only too well, careful not to lose the rhythm. When he finishes a pot, he raises his head from the wheel and heaves a huge sigh.

After attaching handles, spouts, etc., items are left outside the cave to dry in the sun. When a sufficient number of pieces are ready, every two months or so, the kiln is prepared for the wood firing (*cottura*). The kiln is filled and bricked up and, without any form of temperature control, left for three days at around 1832°F (1000°C), with another two or three days for the cooling process, at which point the kiln is opened. Checco Lallo burns mostly olive wood because it creates more heat, and is more readily available.

Checco Lallo produces mainly classic items, although clients ask him to make "modern" objects, such as tea and coffee services or flasks. When he retires an era will end, full stop.

Opposite

Two simple and honest pots with traditional decoration, sprigs of olive and yellow and green flowers. The clay and glaze combination allows these ceramics to withstand direct heat.

Above

A set of darker rustic tableware using traditional colors with manganese.

MARA SANTONI, BUCCHERO, TUSCANY

THE ANCIENT ETRUSCAN Federation loosely controlled the whole of central Italy and examples of their fine pottery exist in most major and minor museums throughout the area. Very few ceramicists, however, dedicate themselves to the very neglected yet beautiful art of *bucchero*, the Etruscan technique of firing clay in a wood kiln to create a black pewter-like effect. Mara Santoni, working in the small hilltop town of Massa Marittima, is one potter who has evolved a range of objects in *bucchero* which maintain the tradition while

"I love the 'magic' of bucchero, *that moment when the clay turns black."*

MARA SANTONI

reinterpretating form. She and her partner, Alessandro Marchetti (see page 131) who works in leather, often pool their talents to produce wonderful plates, bowls, and vases combining *bucchero* and incised leather elements.

Mara learned the *bucchero* technique from an old potter in Massa Marittima some 10 years ago and decided to open her studio soon afterward. The softened clay is rolled out like pastry and shaped in a mold, usually made from plaster of Paris. Once the pot has dried, Mara works the clay by burnishing it with a small spatula made of plastic or bone. Sometimes a layer of slip (liquid clay) is added to give a different texture (i.e. a matte texture on one side and a shimmering texture on the other). The object is wood-fired in a single firing to 1202-1588°F (650-900°C). If the clay is fired at too high a temperature, it turns into the standard burnished red-brown of terracotta. Mara says the "magic" of *bucchero* rests entirely in its color and depth: the kind of black achieved depends on the type of wood used, the more resinous the wood (e.g. pine), the better the silvery sheen; whereas a pure deep black is achieved with sweet chestnut. Only about half an hour of firing is required for a medium-sized bowl.

Above (top and below)

Mara models the clay, first by pressing it over an existing shape and then by finishing the rough form by rubbing and polishing with small implements made of plastic or bone.

Opposite

One of Mara's modern Etruscan inspired ceramic creations that combine *bucchero* and Alessandro Marchetti's etched leather center.

Above

Tiles with hunting scenes designed in the late 1800s on themes from the high Renaissance.

Opposite (clockwise from top left)

A two-handled vase (a *borracia*) from the 17th century with 16th-century decoration; a fruit bowl in a 16th-century shape with 17th-century arabesque decoration including half-woman, half-fish handles; a 20th-century tea set based on 17th-century decoration; a modern ceramic table on a wrought iron base.

UBALDO GRAZIA, CERAMICS, UMBRIA

"We look to our incredible heritage to help us in welcoming new talent to the present and for the future."

UBALDO GRAZIA

A DIFFERENT LEVEL of output comes from the Umbrian town of Deruta which has been at the center of Italian ceramic production for more than 600 years. Deruta has seen it all – a big boom in the 16th century for ceramics with arabesque and figurative designs for which demand collapsed during the following century, followed by a peak in the late 19th century after the reunification of Italy. Today, there are dozens of competing factories throughout the lower, newer town and it is often difficult to distinguish one factory's output from another.

One exception is the small company Grazia, run by Ubaldo Grazia since 1973. Today, roughly half of the 60 workers are occupied in decorating and glazing. From the beginning of the last century, the company has involved painters in the evolution of new designs. At first they were mainly local and Italian but, in the 1970s, American artists started to arrive bringing with them new designs and new color combinations.

Grazia's clay comes from nearby pits or from San Sepulcro. To prepare it for working, it is dissolved in water, then passed through a filtering system to remove impurities. This filtering produces large clay "pancakes" that are then "matured" for six months in a humid environment before being worked. All their production is handmade, either thrown or molded, and biscuit-fired for 36 hours to 1832°F (1000 °C). After decorating and glazing, items are fired again to around 1724°F (940°C).

Grazia only produce classic *maiolica* but in every style from the 1400s onwards, including the very latest in modern decorative styles. They are also the only factory that controls the complete output of its products from start to finish.

Right

Two examples of folklore statuettes or *figurine* made by Maurizio Patrì which were very popular in the 19th century when the figures were originally dressed with real textiles. On the left is a cheese vendor and on the right a gypsy girl who tells fortunes by getting the bird to pick one of the tarot-cards she keeps in a box.

Opposite

Ceramic pots in the form of heads date back to the 17th century and are particular to Sicily.

44

MAURIZIO PATRÌ, TRADITIONAL CERAMICS, SICILY

"Modern day Caltagirone is awash with ceramic botteghe. We are one of the few who respect the traditions handed down from our father Maurizio and his father who were both important ceramicists here."

MARCO PATRÌ

THE TOWN OF CALTAGIRONE, whose name derives from the Arabic *qal'at'* and *ghirân* (hill or rock of vases), had produced ceramics since the early Arab period and had become particularly famous for its magnificent polychrome floors. The 19th century was particularly important for the evolution of the figurine or folklore statuettes and decorative oil lamps in the form of male and female characters.

Modern-day Caltagirone is packed with ceramic *botteghe* (workshops), and there is a constantly renewed work force from the 100 or so students who attend the well-established School of Ceramics there. Most of the ceramicists working, however, do so as a way of making a living without necessarily respecting the centuries-old traditions of the town. The Patrì family are one of the few from the 140 *botteghe* who have remained faithful to the past.

The brothers Marco and Màrcello Patrì run a small shop in the center of town. Their father Maurizio and grandfather were both important ceramicists in Caltagirone in the days when they used to make their own clays from local deposits. The laborious process consisted of digging the clay out of the riverbeds, drying it out in the sun, then breaking the lumps down into manageable pieces to be transported by donkey to their *botteghe*. The dried pieces of clay were placed in vats of water, sieved to remove debris, and reworked to the correct consistency, then left in a damp storeroom until required. Although modern clay production techniques are less arduous, the brothers still acquire their own raw materials and never use commercially produced clays.

Maurizio, who began painting ceramics by the time he was six, enrolled in the ceramics school. He met his wife there and sent their three sons there in due course. Marco now specializes in decorative ceramics that are either evolutions of local styles or improvizations on other Italian designs, whereas Marcello has developed an antique style of painting and distressing.

45

BOTTEGA NICOLA GIUSTINIANI, 18TH-CENTURY CERAMICS, CAMPANIA

"From the earth, from water, out of the fire and in expert hands, an object is born, symbol of a 18th-century tradition renewed every day in forms and color at the Bottega Nicola Giustiniani."

ELVIO SAGNELLA

EARTHQUAKES ARE AN ever-present threat in Italy and many have been so devastating that whole regions have been severely disadvantaged for decades, sometimes centuries afterward. In other situations, an earthquake has actually resulted in the revitalization of the area destroyed. This happened in an area east of Naples, near Benevento in Campania, after the earthquake of 1668.

Two towns, Cerreto Sannita and nearby San Lorenzello, had developed at the same time. Sitting higher than San Lorenzello and built on a clay spur, Cerreto was virtually destroyed in the 1668 earthquake, whereas lower-lying San Lorenzello suffered much less. The rebuilding of Cerreto helped to revive the artisan workshops of San Lorenzello, including some of those displaced from Cerreto. Others arrived from Naples, including the ceramicist Antonio Giustiniani. His son Nicola, born in 1732, became one of the most influential figures in Neapolitan ceramics history.

The arrival of a new Spanish King at the court of Naples (which stimulated the demand for refined goods), the abundance of clay, and a profusion of established workers in San Lorenzello provided a climate for the Giustiniani family workshop to produce ceramics of ever-higher quality, at times approaching porcelain levels of sophistication. They married into the Feste family around 1775 and continued to produce excellent ceramics until the end of the Napoleonic period in Italy.

More recently, San Lorenzello has gone through another renaissance under the influence of the dedicated artisan potter Elvio Sagnella, who opened his Bottega Nicola Giustiniani in 1981. A graduate of the nearby Cerreto Sannita Art Institute, Elvio has taken the traditions of the past, as exemplified by the works of Antonio and Nicola Giustiniani, and brought them back to life in both old and new forms. Their workshop still specializes in traditional objects decorated in the same local colors as in the past – yellows and greens and especially the blues for their Neapolitan customers.

Opposite (far left)

This elaborately designed and colorful decorative vase evolved from a storage bottle used in the 19th century.

Opposite (top right)

Some of Bottega Nicola Giustiniani's range of plates. Five traditional local colors are usually used. Yellow and green (magnesium and copper oxide) are very popular with blue (cobalt) mostly preferred by customers from Naples.

Opposite (below right)

A large "Pulcinella" plate based on traditional Commedia dell'Arte scenes.

47

DEL MONACO, TRADITIONAL PUGLIAN CERAMICS, PUGLIA

Right

A small simply decorated jug whose shape and form were well established in Puglia at least 600 years ago.

Opposite

This elaborately decorated acorn exemplifies the richness of the Puglian heritage and, in particular, reflects the traditions brought from Sicily by the 13th-century influx of Arab-Sicilian artisans whose geometric-abstract designs were conditioned by religious restrictions. It is a decorative piece; it has no practical function.

GROTTAGLIE, ON THE INSTEP of Italy's heel near Taranto in Puglia, was probably founded by immigrant Greeks and may have been an important ceramic center in the ancient past. The 13th-century influx of Arab-Sicilian artisans into Puglia revitalized ceramic production, and one of the oldest established and most respected families still working there can trace their origins back 600 years.

In 1416 Maestro Leonardo del Monaco established a small factory in an area of Grottaglie dedicated to artisans. Over the centuries, La Bottega del Monaco has passed the skills and secrets of its craft to each successive generation without compromising the quality of its products while, at the same time, maintaining local values.

Vincenzo del Monaco (c1925–1978) dedicated himself to a total revision of the technical and productive capacities of the family business in parallel

"We seek to maintain local values without compromising the quality of our products. This is the legacy that our father gave us and by which we live."

ORAZIO DEL MONACO

with painstaking research into the materials and designs of the past. Slowly, he and his sons, Orazio and Giuseppe, expanded and refined their ceramic output to arrive at the point where, today, the family is recognized, nationally and internationally, as one of the best producers of traditional Grottaglie ceramics.

Maestro Vincenzo also passionately supported the local "Istituto d'Arte per la Ceramica," the oldest ceramic institute in Puglia (founded in 1887), and through his dedication and inspiration helped many of the current generation of working ceramicists. Orazio and Giuseppe work with their sister Elisabeta in their small workshop carved out of the rock, surrounded by many other producers, in the same part of Grottaglie that artisans have inhabited for perhaps 2,000 years.

A very expressive polychrome terracotta Madonna and Child whose stylistic inspiration comes from the medieval paintings of the Florentine Cimabue.

"In terms of form and technique, I believe I am a better craftsman than my father but he had a passion and inspiration which made his work very special and which gave it a deeply emotional dimension. One day I hope to match this aspect of his work."

LUCA VANNI

50

Right

A simple Madonna and Child and the mold from which it is made.

Far right

A detail of one of Luca's dazzling terracotta cityscapes of Florence, the city that has always had a powerful influence over him and his home town of Impruneta.

LUCA VANNI, TERRACOTTA, TUSCANY

IN PARALLEL WITH the arrival of *maiolica* tableware in the form of plates, pots. and vases, terracotta objects started to reappear in Italy, particularly in molded form and especially on the façades and portals of buildings. Noted artists such as Donatello and Jacopo della Quercia produced relief sculptures in terracotta while the Florentine, Luca della Robbia, executed brilliantly enameled plaques and medallions and raised this form of terracotta from a craft to high art. Despite this wonderful flowering of three-dimensional glazed terracotta art, plain terracotta is now perceived as a functional material, used for simple or decorative garden pots and planters or for the "*orcio,*" the large storage jar used since antiquity to store olive oil.

The town of Impruneta, south of Florence, is known for its terracotta production, primarily because the local clay is naturally frost resistant. Although much of the local output is of good quality, it tends toward very traditional, conservative styles. Luca Vanni, born and raised in Impruneta, is one exception. Luca's father, Antonio, had modeled terracotta as a hobby from boyhood and, over the years, developed a local reputation

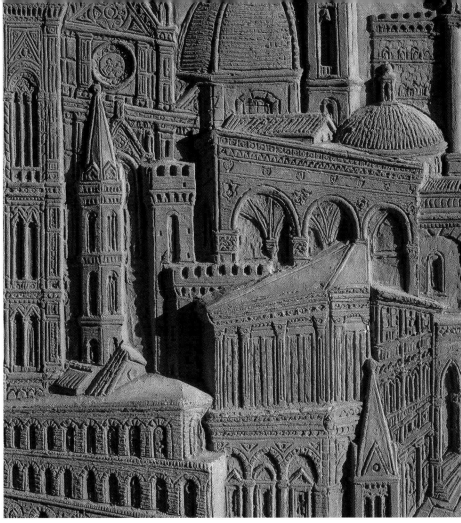

for his religious frieze work. Gradually, Luca started to take an interest in his father's work. For two or three years he worked alongside Antonio and then started a workshop of his own. An important opportunity came when Antonio got a commission from a monastery for an arched frieze or panel that he passed on to Luca to execute.

This first piece launched his career, and other important commissions followed, some of which are now housed in the small museum in Impruneta, including one of his most famous pieces, a huge Christ on the Cross.

All Luca's output requires the creation of a mold from which he then creates the finished terracotta object. The firing of the clay is carried out in an electric kiln at precisely 1736°F (947°C). The molds are taken out, allowed to cool a little, and then soaked in water for a couple of hours to halt any further chemical processes which are activated during the firing. Quality terracotta has a clear bell-like sound when struck firmly; terracotta that is not so well fired (particularly items that sit at the bottom of the kiln, for example) produces a much duller sound.

Luca spends most of his time making elegant unglazed terracotta objects for the market – bird plaques, friezes, Madonnas, and the like – in his very tidy workshop on the edge of Impruneta. His wife paints the plaques and figures, and his brother helps him occasionally. Their works are sold in shops in Nagoya (Japan), Rome, Impruneta, Florence, and Milan.

wood ~ the tree of life

The many thousands of tree varieties provide woods of specific characteristics, all unique, combining in differing degrees, hardness, elasticity, stability, texture, color, grain, and even scent. Oak is known for its strength and durability and was used very early on in timber frame construction and shipbuilding; ash for tool handles because of its toughness; the close grain, color, and hardness of walnut and mahogany are valued for furniture; boxwood and lime are preferred for very fine detail work; and pear was once used in printing and other forms of block work because of its hardness and stability. Little historical evidence of wood used in the home exists because wood deteriorates much more quickly than its competitor materials. However, it is probable that the Egyptians first used lathes to make chair legs, and the Romans invented a form of the plane. The Romans may have also been the first to use iron nails, mainly in holding together large wooden constructions. Renaissance structural woodwork was essentially an imitation of architecture, with pilasters and entablatures being particularly favored. In his Urbino Palazzo Ducale, Duke Federico's study walls were covered with splendid intarsia work by Botticelli and Baccio Pontelli, demonstrating the exceptional pictorial qualities that fine artisans could bring to wood paneling. Marquetry, essentially a wooden version of stone inlay work, has been the only major innovation in furniture making since Renaissance times. Floral, arabesque, or figurative designs formed of thin pieces of exotic or colored woods, originally cut with a hand saw, are glued directly to the furniture surface.

opposite: Valter Rottin's wooden bottles.

LORENZO PACINI, REPRODUCTION FURNITURE, TUSCANY

Motto "Di solo non si fa niente" – *alone little can be achieved.*

LORENZO PACINI

A TRUE TUSCAN from Lucca, Lorenzo Pacini's father was a state functionary who made furniture in his spare time for the family home and occasionally for friends. In his late teens, Lorenzo decided to make furniture on his own account – with just one old artisan carpenter, a little money, and his innate aesthetic taste. Lorenzo's spirit and family values went against the 1970s tendencies in furniture making especially that of the *povera* style (simple and rough). Instead he concentrated on producing only immaculate reproductions of 15th-century Renaissance Italian furniture.

As demand for his designs increased, the space in the garage underneath their home proved insufficient, so he persuaded his mother to move the vegetable garden to another part of their small property on the outskirts of Lucca and built in its place a larger workshop to accommodate other artisans.

The only way a small manufacturer could make an impact on the market was to go to the furniture fairs whenever possible, especially to the international fair in Milan. In the early years, the main problem for Lorenzo and his young wife Rosangela was in explaining products of quality to potential clients. However, he persisted by ensuring that his stand was the most innovative and stylish possible, using fresh flowers from the countryside or devising interiors using "false" materials such as painted polystyrene tiles to fake real ones. In the early years they made many sacrifices to keep the business afloat, working Saturdays and Sundays, visiting clients to measure rooms and even selling their matrimonial bedroom furniture, not once but three times.

They were becoming desperate for a substantial commission when they had their big break at the 1982 Milan fair. A lady from Jordan, completely unknown to them, ordered Lorenzo's "father's room" for the son of the then King Hussein. Lorenzo and

Left

Restored 18th-century Louis XVI chest of drawers using fine marquetry or *intaglio*. Lorenzo Pacini discovered one highly skilled worker in Florence in the 1980s who was able to bring intaglio techniques to their furniture range.

Opposite (above)

17th-century reproduction Baroque Italian sideboard using *tempera antica* paint effect for the decoration.

Opposite (below)

Detail of the effect achieved by fine *tempera antica* on a "Liberty" wardrobe, early 20th century.

55

a fitter went to Jordan to install this "room," and they came back with more orders from the Jordanian Royal family.

Their early pieces primarily used local chestnut (*castagno*). As they developed 1700s and 1800s furniture lines, more walnut (*noce*) was introduced. Later cherry (*ciliegio*) and other rarer woods were included in the production of special pieces. Virtually all the Pacini wood comes from Italian sources and much of that from the Garfagnana and Versilia areas north of Lucca, where forests of chestnut and walnut are still plentiful. All of their furniture is made from solid timber (*masello*) – they never use veneers. Lorenzo is so careful about his timber sources that he visits the forests to see exactly where his trees come from. For him, the location of the tree and time of cutting are as important as the right amount of seasoning necessary before choosing the timber.

At the end of the 1980s, Lorenzo came across three exceptional artisans in Florence who understood the use of *tempera antica* effect on wood, essentially consisting of the application of a thin layer of gesso (plaster) which is then stressed and decorated in various ways. Around the same time he discovered another highly skilled worker able to bring intaglio designs to their furniture range.

Lorenzo Pacini views his world as a collaboration between himself, the designers, and the artisans without whose skills he could not have developed the business. His own skill is the ability to visualize any project in his head, from one piece to a whole room set. It might appear that Lorenzo Pacini's painstakingly made furniture comes with a hefty price tag but this is not so. Their artisan-made furniture will stand the great test of time which cannot always be said of the factory-made piece.

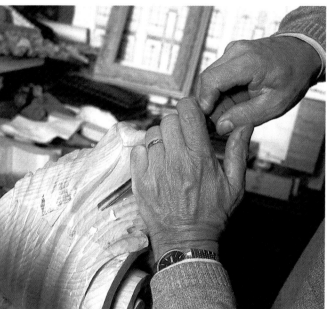

Top & Above

One of the Bartolozzi e Maioli craftmen has been working here for 45 years – since the age of 16. He makes everything from small wall-mounted light frames to 20-foot (6-m) high carved wooden doors utilizing any number of the 300 or so tools on and around his workbench.

LA BOTTEGA D'ARTE BARTOLOZZI E MAIOLI, WOODCARVING AND RESTORATION, TUSCANY

"Bartolozzi e Maioli is a direct descendant of a typical Florentine Renaissance workshop, operating within the field of applied arts, and I am determined to ensure that these crafts continue to live."

FIORENZO BARTOLOZZI

FIORENZO BARTOLOZZI, from Florence, and Giuseppe Maioli, from Ravenna, shared a passion for the art of carving wood. From a casual encounter, an intense artistic collaboration started, out of which La Bottega d'Arte Bartolozzi e Maioli was born in 1938.

From 1940 to 1944, the workshop ceased functioning due to the war, but as soon as it ended, they were immediately involved in the restoration of all the wooden elements of the sacristy and choir of the heavily bombed Abbey of Montecassino, between Rome and Naples. Over the 12 years that this project lasted, a spontaneous school of apprenticeship grew at Montecassino with the two *maestri* (masters) at its apex.

Little by little, and as their names circulated within the world of restoration, their *bottega* in Florence became a cultural and artistic meeting point where painters, actors, directors, singers, and others from the artistic community mixed with bankers, doctors, and politicians. Giuseppe and Fiorenzo's experience and skill combined well with their growing portfolio of contacts. Over the years they have worked on many prestigious projects from the Quirinale in Rome (the Italian Parliament) to a sculpture of Christ in Dallas Cathedral. The *bottega* has also been involved in the restoration of two large state rooms in the Kremlin, in Moscow.

At the same time as carrying out high-profile restorations, Giuseppe and Fiorenzo collected fragments, architectural pieces, wooden statues, picture frames – indeed anything old, made of wood. Today, this collection representing the minor arts is

Left

This circular wooden block will become a ceiling rose, part of a major Kremlin restoration project which Bartolozzi e Maioli are carrying out for the Russian government. The original ceiling rose is in the background.

Overleaf

The Bartolozzi e Maioli store rooms, are piled high with frames, chairs, tables, giant statues, and sculptures. The various depositories are packed with wooden objects of all shapes, sizes, and from all eras. As well as making new artifacts based on this incredible legacy, they have a huge restoration business.

possibly unique in the world and is essentially a museum of Italian art history, which assists in keeping these arts alive and permits the *bottega* to re-create any wooden object imaginable.

This amazing collection, spread around a number of workshops and "showrooms" near the Pitti Palace in Florence, itself an area where many artisans have always plied their trade, allows the new to be compared with the old in order to correct errors of style or technique before they are finished. La Bottega d'Arte Bartolozzi e Maioli is continuing the traditions of a typical Florentine Renaissance workshop, translating ideas through craftsmanship into beautiful artifacts.

ANTONIO IACHINI, MARQUETRY AND RESTORATION, LAZIO

ANTONIO IACHINI was born in Tuscania, a small medieval town in the northern part of Lazio known for its Etruscan and Romanesque roots. On February 6, 1971, a powerful earthquake hit Tuscania and the surrounding countryside, bringing to light, literally, *pozzi* (wells) that had lain undiscovered for centuries. At the bottom of these wells excavations revealed layers of ceramics including many medieval pieces. Antonio, a "natural" designer, who had already begun to master fresco restoration through drawing, reading, and experience, soon applied himself to the restoration and decoration of these found ceramics.

Still a single man with a great desire to learn, he decided to leave Tuscania to enlarge his experience and went to Basilicata where a friend in Matera offered him an apprenticeship. For two years he worked on fresco restoration and greatly enjoyed the experience. But his greater desire was to become independent. It was at this point that he started to work with wood and immediately felt that this was his material. Antonio would often go to the woods to cut timber just to understand more about the basic material. His early experiences in wood involved restoring furniture for antique dealers. In 1980 Antonio experienced another earthquake, this time in Matera. He felt somehow connected with these events, and he approached Michele Dilia, the

Top left

Antonio carving a wooden rose. He is particularly known for his restoration of old ecclesiastical "furniture" pieces.

Top right

Antonio Iachini is very proud of this "trompe l'oeil" intarsia composition, which he completed after his work experiences at the Palazzo Ducale in Urbino.

Sovrintendenza (Superintendent of Restoration), someone who had always believed in and encouraged Antonio, and offered to help with his small lorry. He thus started his first tentative involvement with the state by recovering broken material for the superintendent. To further his education, Antonio attended UNESCO conferences on restoration and the *Collegio Inginiere della Toscana* in Florence to increase his knowledge and to learn new techniques.

Antonio left Matera in 1987, after he had married a local girl, to return to Tuscania where he opened his workshop. He returned occasionally to work in Matera but these commissions came to an end after a change in the *Sovrintendenza* there and the offer of work in nearer Urbino (Le Marche).

His love for intarsia came out of his experiences at the Palazzo Ducale in Urbino where Duke Federico's study is celebrated for its walls, entirely composed of wood intarsia by Botticelli and Pontelli and acknowledged as the height of Renaissance achievement in this art. Initially Antonio felt very humble in front of these works but felt that he could achieve their skills with dedication and time.

In his restoration, Antonio puts himself in the shoes of the creator of the piece and tries to "become" the artist by entering into his "spirit," by looking, taking photographs, and by working out what was involved; and, much more importantly, he asks himself why the artist created the piece in that particular way. Once all this has been considered, Antonio then works to re-create the original vision.

"I now feel less afraid in front of the great masters of the past and feel able to approach them on a more equal footing but never without great, great respect for all those who preceded me."

ANTONIO IACHINI

The Duomo of Ascoli Piceno provided Antonio with an interesting challenge. He first studied the wooden gothic choir stalls some 10 years before and had noticed that a newer central section was out of keeping with the flanking "original" pieces. When he was given the opportunity of restoring it to its late medieval design, it gave Antonio the stimulus to learn enough to be able to "confront" the original choir designer of the stalls. In fact, through his work and observations, Antonio has been able to rewrite the history of the stalls.

Antonio's small, tidy workshop in Tuscania employs three or four workers whose main tasks include the restoration of furniture for antique dealers and some private clients. But he stresses quite firmly that they have the ability to make any piece of furniture "new," from a chair to a chest of drawers to a large wardrobe. He also stresses the absolute need for respect for the work to be carried out. Much of the basic material for restoring old pieces or unusual wood requirements comes from a special dealer in Macerata (Le Marche). Antonio is proud of his store of at least 100 different timbers and just loves to appreciate their differences and is, as he happily admits, obsessive about them — a truly great artisan!

VALTER ROTTIN, DECORATIVE PIECES AND FURNITURE, VENETO

"The inspiration for the bottle forms grew naturally from the shapes and grain I see in the wood, and I really enjoy juxtaposing different woods to create interesting contrasts of texture and color."

VALTER ROTTIN

TREVISO, IN THE NORTH of Italy, is one of the Veneto's richest provincial capitals and was for four centuries a loyal ally of nearby Venice. Although not particularly known for artisans working in wood, there is a long tradition in most of the Veneto of furniture manufacture utilizing timber from Istria and other areas of the former Yugoslavia.

61

Valter Rottin works on his own in somewhat less than salubrious surroundings. His workshop, under an overpass near Treviso's railway station, is a chaos of wood and furniture in the making. A thick layer of wood dust covers old and battered tools and lathes. His small office and "showroom" display his bottles and vases on shelves in a rather higgledy-piggledy fashion.

Valter is endearingly passionate about wood. Even as a child he loved the feel, smell, shapes, knots, lines, and textures of wood. He says that he doesn't know where his passion comes from apart from *"dentro di*

Left

Two contrasting pieces of burr walnut, one natural, the other polished.

Below

After gluing the different wood types together into a solid block, Valter decides what shape to create.

me" – from somewhere deep inside. There is not even a tradition of woodworking in his family, although his father was a craftsman repairing leather car seats.

He studied at the *Falegname Polivalente* for two years, a technical institute in Treviso, where general manual skills are taught in various craft-based areas, specializing for a further year in wood and cabinetmaking. An avid learner, Valter was not easily satisfied and sought to stretch and challenge the knowledge he was acquiring through formal training. Through talking to and working with older craftspeople and reading about the old and traditional, he learned to make furniture exactly as it was made in the 17th century, including a most remarkable antique six-door wardrobe he made for his bedroom at home which took him about one month to make.

Valter buys leftovers that furniture manufacturers throw away – the timber near the top of the tree and root parts, because it is from these that the grain and growth pattern can be understood. For Valter, the most fascinating part is where the main branches leave the trunk. He calls this the "jewel in the crown" because it contains a dense wavelike grain structure that can be used creatively and decoratively in many ways.

Valter's "ordinary" furniture and kitchen cabinets are superb pieces of craftsmanship. However simple they appear to the casual onlooker, when he explains that he only used single pieces of wood for a door (manufacturers piece parts together), to ensure that the panels have similar grain and patterning, the furniture takes on a whole new meaning. He is very selective in purchasing timber, always allowing it to season naturally before working with it, thus avoiding problems such as shrinking, splitting, or bowing later on. He also uses mostly European timber, mainly for cost reasons because timber from further afield is generally much more expensive and difficult to obtain with the qualities that he prefers.

Valter started making his wooden bottles series about 10 years ago, partly because of his great love for the natural colors of different types of wood. These solid shapes are formed as he carves away the waste with a rasp until he achieves a rough approximate of the finished item.

It is so clear that Valter just loves wood and anything to do with wood. He not only makes furniture and bottles but abstract sculpture as well, often using his spare time to develop new ideas. He also only makes furniture that he likes because he just cannot work on a piece if he doesn't like it – not even for the business. Working this way doesn't make Valter financially very successful but it does make him a very contented man whose heart is as big as his personality is quiet.

Right

Two of Valter's solid wooden vases demonstrate the range of colorings, textures, and contrasts that can be achieved by using a wide variety of woods. The vases are resting on a table that has a strip of square wooden mosaic tesserae embedded in the center.

PINO VALENTI, INTARSIA AND MOSAICS, SICILY

Above

Coppia Utopica
(A Couple in Utopia,
1997) in *tarsia* (short
for *intarsia*) and
mosaic, 73 x 49 in.
(183 x 123 cm).
Shows a couple
absorbing the
different colours and
shapes of all the
people in the world
around them.

"In my work I am inspired by many things but especially music, my children, and political issues. I also like to feature letters and numbers in my pictures to add to the narrative."

PINO VALENTI

PINO VALENTI WORKS at the opposite end of Italy to Valter Rottin. In their respective early experiences, formal training and first years of work, they have much in common, but Pino has undergone a further transformation and has evolved from a dedicated artisan into a passionate artist by using all his skills to create highly original pictures in wood – not as commissions, nor even to satisfy some kind of market.

Pino and his family live in a terraced townhouse on four levels in the small town of Collesano, a short distance inland from the historic Norman city of Cefalù on the north Sicilian coast. Everything wooden in the house has been made by Pino – all the kitchen units (even a housing for the microwave), the doors, the tables, the sofas, the desk, and the beds and wardrobes, too.

Left

A detail of an unfinished work in *tarsia* showing blocks of wood weighing heavily on "thoughts, solidarity, democracy, etc." A particularly personal statement on how an artist copes with the specific demands and needs of his family and how these relate to the rest of his responsibilities to the world.

Below

Pino working on the details of a table commissioned by the *Collesano Municipio* (Town Hall) in commemoration of victims of the Mafia including Della Chiesa, Falcone, and Borsellino. The tableau is entirely Pino's and demonstrates not only his stong anti-Mafia commitment but that of the town as well.

65

Right

Barca di Carta (Boat of Paper, 1994) in *tarsia* and mosaic, 69 x 49 in. (173 x 123 cm). A visually
simple image of a couple sailing away on a boat with paper sails except that the sails are made of one of Sicily's leading newspapers and relate to political events understood by many Sicilians.

Pino was born and raised in Collesano and his family still lives there; his wife is from another small town nearby. When he first started to earn his living, he worked in a carpentry shop making standard windows, doors, and the like. But apart from learning certain skills, it ultimately bored Pino, so he studied arts and crafts at the Institute of Arts in Cefalù, where they teach a wide range of subjects including weaving, wrought-iron work, and woodwork. It was, and still is, a three-year course but, when Pino studied there, they taught practical skills; today theory is considered more relevant. Pino strongly questions this approach to education for he sees it as "... no longer preserving and celebrating through practical application something historic and traditional, rather it is something being seen as a museum object...." In other words, theory without practical implementation is worse than useless.

After this period of formal education, Pino set up on his own as a general carpenter still making windows and doors but also furniture to his own designs. One day, after 10 quite successful years, a Milanese architect friend asked him to make a replica of a painting, but in wood. It was from Picasso's blue period and Pino rightly asked: "Where do you get blue wood from?" With his friend's help they found a company near Milan that specializes in thin sheets of stained wood veneers. Having completed the "Picasso" for his friend, he found that he had all these little bits of wood left over, so he started to play around with shapes and colors.

Above left

Part of *Ragazze al Mare* (Girls at the Seaside, 1997) in *tarsia* and mosaic, 49 x 49 in.(123 x 123 cm). Two girls on the beach depicted in simple lines, colors and forms. This composition shows calmness and tranquility.

Above right

Cavalli (Horses, 1990) wood on *tarsia*. An early work of dynamic expressiveness, using contrasting woods and shapes.

At first he copied existing paintings of Michelangelo or Botticelli which, in their own way, are quite remarkable, but gradually he began to develop his own style to arrive at his own extraordinary compositions, full of color and expression. Pino's methodology is quite basic. First he makes a drawing to scale, incorporating his design into what appears on the page as a jigsaw. He cuts and shapes different colored veneers with a sharp chisel to form the main subject and then, by using tiny squares of wood in mosaic, creates a different, contrasting texture (a kind of Seurat effect) for the background. The mosaic provides a three-dimensional effect and allows for a greater degree of shading and depth within the composition. Pino happily mixes his own version of wood relief, veneer marquetry, and mosaic in the design, according to his vision of the panel.

He also incorporates all manner of natural woods, often with differing thicknesses to provide depth, texture, and color. Interestingly, the bright yellow wood he sometimes uses actually is lemon tree wood.

67

First impressions of his work are of simple and bold designs with great and immediate impact. However, his figures are extremely expressive both in their facial expressions and in their bodies. Often there is also a great deal of movement, not just in the twists and turns of the bodies or their striking poses, but in the sense that they are trying to break out of the frame. Other panels display great calmness and tranquillity.

His sources of inspiration are diverse: from music (several pieces are his interpretation of songs by the great Italian troubadour, songwriter-poet Paolo Contc), to his children, to political issues. He likes to feature letters and numbers in his pictures, mostly with references to "the meaning of life" when a darker side to the narrative is required, but generally he seems to have a positive and upbeat approach to his work, including a wry sense of humor and a certain Italian sentiment.

Occasionally he will work on a commission if the project is important to him personally. For example, he has recently completed a table for a special room in the *Collesano Municipio* (Town Hall) in commemoration of victims of the Mafia, demonstrating his support for the cause.

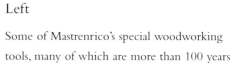

Left

Some of Mastrenrico's special woodworking tools, many of which are more than 100 years old and are irreplaceable.

Above

Mastrenrico using a 19th-century hand-plane to hollow out a milk pail made out of solid wood (any wooden container intended to hold liquid cannot have seams).

Right

Mastrenrico finishes chairs with a knife to bring out the grain of the wood, whereas a machine-plane would just flatten it.

Far Right

Detail of the grain brought out by this process.

MASTRENRICO, UTILITY WARE, VENETO

"This is our craft, and it is what we believe in."

ENRICO DI PASAVENTO

THE VERY EXISTENCE OF this small family business situated just outside the medieval walls of Montagnana, between Vicenza, Ferrara, and Padua, initially comes as a great surprise. However, the combination of a resurgence of interest in organic, natural, handmade goods and the Veneto being one of the richest areas of Europe, never mind Italy, allows Mastrenrico to make and sell his handmade wares from this location. The family ancestors came originally from the mountains around Asiago, the Alto Piano dei Sette Comuni, north of Vicenza. Traditionally, mountain peoples spent the winter months making wooden tools and domestic items to bring down to the plains below in spring. Mastrenrico's grandfather was a wicker specialist and received a diploma in 1910 for the best work in willow (*vimine*).

Mastrenrico, Enrico di Pasavento is his given name, has been making the same range of wooden items as his grandfather's family made since he was a boy. His wife makes rush seating for the chairs, and now their son, without any pressure from his parents, is happily following in his father's footsteps. They make a massive range of items, using many tools that have been handed down from generation to generation. Their range of items includes furniture such as wooden chairs, rocking chairs, and basket-weave cribs; kitchen utensils and household items - wooden spoons, forks, slicers, graters, ladles, lemon squeezers, pasta rollers, scoops, spatulas, traditional wine stops, linen baskets, trays (there is a growing market for these items not just for their functionality but also for their decorative value in the kitchen); and traditional working items such as small harnesses, goat collars, churns and pails, and practical baskets, such as the one made for gathering mushrooms which is specially curved to sit comfortably against a woman's hip.

69

Above

A *gerla* or wicker basket, made for carrying wood on one's back, is still used in some alpine regions of Italy. Mastrenrico both restores them – he has one which dates back 100 years – and makes them.

OSVALDO SANTILLO, BASKETS, CAMPANIA

"More than anything, I look forward to the Christmas feste *when I can make special cribs for the little church nearby. One year was very special because my baby grandchildren played the role of 'Jesus in the manger' in one of my creations."*

OSVALDO SANTILLO

THE LONG EXISTENCE of many crafts is often attributable to the fact that the population was isolated and less subject to outside influences, as in Sardinia, where each region is represented by a different style of basket according to the materials sourced in the immediate locality. In southern Italy, in similar situations of isolation, the evolution of crafts is the same.

Basket-weaver Osvaldo Santillo sits most days on the terrace outside his home in the small hilltop village of Pietrelcina near Benevento where, with a combination of patience and passion, he makes simple household items from the cane and willow he collects from the surrounding countryside. He was born there and has always lived there. He works steadily, untouched by the concerns of the world at large. For 30 years Osvaldo has worked alone and has no one to whom he can pass on his skill. Unlike Mastrenrico in the north (see pages 68–69), his family is not interested in perpetuating this ancient craft.

Basket weaving is one of the oldest handcrafts known to man and represents the earliest use of plant fiber. Its origins have always been practical, for making containers for use in the home or in the field, and the craft flourishes in any rural area where there is a plentiful supply of raw materials – reeds, bamboo, palms, grasses. Osvaldo's raw materials are primarily cane, willow and straw, and his baskets are very simple, unadorned objects. His baskets range from small table varieties to larger hampers, shopping and laundry baskets.

Opposite (above)

Stacks of raw materials collected locally by Osvaldo, where there is a plentiful supply: reeds, bamboo, palms, grasses, canes, willows, and straw.

Opposite (below)

Osvaldo Santillo working on one of his baskets outside the workshop near his home in the small hilltop village of Pietrelcina near Benevento, Campania.

Left

The vertical structure of this particular basket is made of cane, through which Osvaldo interweaves willow that has been soaked in water for about 10 days to make it pliable.

71

metal ~ forged by passion

The Villanovans, predecessors of the Etruscans, were the first people in central Italy to use iron. They, the Etruscans and, later, the Romans worked the copper and iron mines, particularly those of Tuscany and Elba. All produced fine metalwork of varying designs, often influenced by Greek art. By the late medieval period, iron ornament in scrolls and arabesques covered church doors. In Italy in the 13th century, the Venetians introduced screens pierced with geometric patterns cut into unheated metal. However, the processes involved in metallurgy moved slowly until about 1300 when advances were made, first in Spain where the Catalan forge allowed a large quantity of iron to be produced at one time, and then in Germany where cast-iron techniques developed. Until the middle of the 19th century, wrought iron used structurally and decoratively predominated over cast iron because of its cost advantages. Wrought iron can be hammered, twisted, and stretched when hot or cold and made into a vast variety of decorative objects as well as agricultural, industrial, and military implements. The Industrial Revolution transformed the character of the metal industries when cast iron was mass-produced for the first time and the art of the decorative smith declined. Particularly interesting is the town of Agnone, in southern Italy, which became an economic center of great importance, strategically situated midway between the Tyrrhenian and Adriatic seas where bronze bell casting and copper manufacturing flourished.

opposite: Rafaele di Prinzio: wrought-iron head boards showing the marks of the hammer blows, which symbolically represent both the force contained within as well as the imprint of the artisan at work.

74

Above

The hand-built bell core, a brick construction corresponding to the interior part of the bell, prepared with a wooden mold.

Opposite

A "Jubilaeum" bell weighing about 2,200 lb. (1,000 kg). Each bell is made individually, and each bell's tone is different. Bell makers talk of a bell's
soul and how the sound is expressed through the bronze.

FONDERIA MARINELLI, BELLS, MOLISE

"The Marinelli family has outlived all of the other bell casters of Agnone and their traditions of casting have been transmitted from father to son over succeeding generations."

FONDERIA MARINELLI

THE PERIOD UP until the late 18th century, despite many changes in feudal ownership, saw Agnone develop as an economic center of great importance in this part of Italy. For centuries, the two most important activities, other than livestock, which dominated the town's output were copper production and bell making. The former is no longer the industry that it once was but the latter is still successfully carried on by the dedicated Marinelli family.

The importance of bells in the story of Christianity is slowly becoming a distant memory. However, in the 5th century and not long after the Roman Emperor Constantine's decree that allowed Christians to practice openly, Saint Paolino of Nola in Campania (derived from the Latin for bell, *campana*) introduced the use of bells into the church liturgy. Their symbolic and functional importance was underlined in the 8th century, when the blessing of a new bell by the local bishop came into being and, as the building boom of the new millennium accelerated, bell towers became significant elements of the new churches.

The Marinelli Foundry is possibly the oldest bell foundry still functioning in the Western world. Family members may have been working in this field by the year 1000. The Marinelli family has outlived all the other bell casters of Agnone and has kept alive the traditional methods of casting.

Casting a bell in bronze, specifically 78 parts copper and 22 parts tin, is still considered a work of art at the Marinelli Foundry, with each bell having a character all of its own. The form, decorations, and tone are always specific to the year it was made

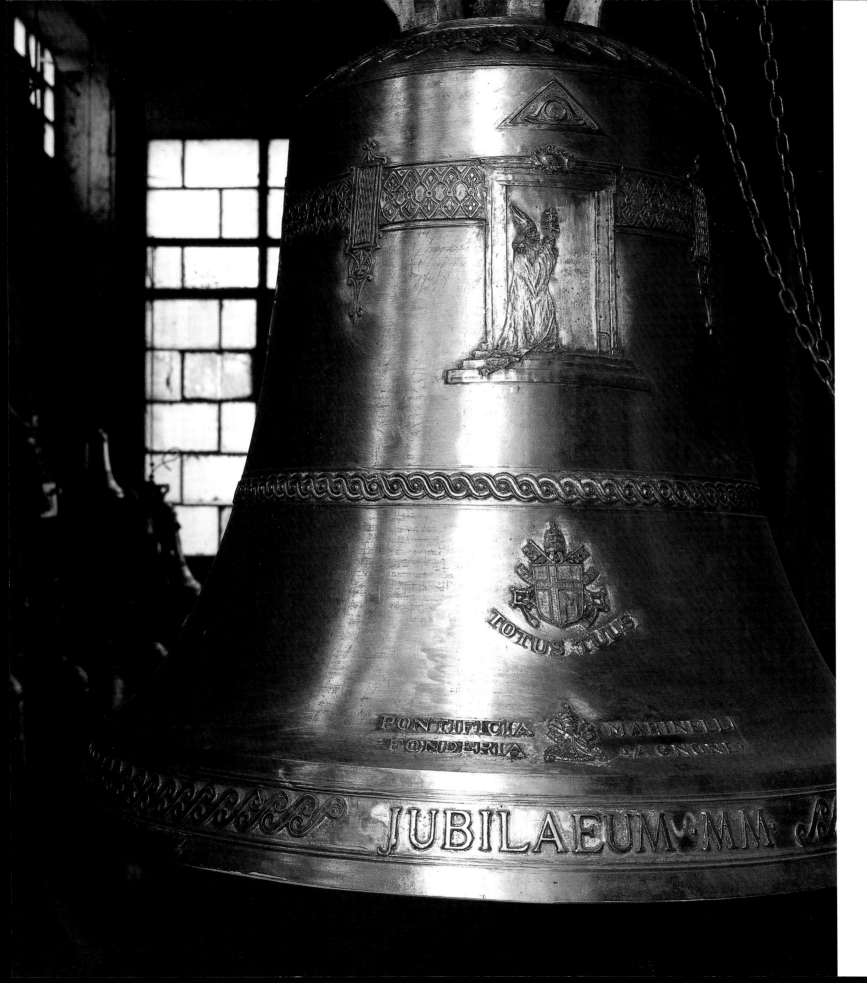

and to the caster. Over the course of time, decorations have evolved, from simple designs outlining symbols, coats of arms, and seals, to more complex images of the Virgin and Child, archangels, crosses, and crucifixes. The inscriptions, too, have varied enormously – from prayers and messages, to names of the donor or the priests or even just of the caster.

Until the early 16th century, bell making had hardly changed for 500 years. But bells became overdecorated, affecting tone substantially. The application of science in the 16th century changed the way bells were cast technically, with tone being the dominant factor in bell casting and decoration being kept prudently minimal.

The actual manufacturing process is long and laborious. The crucial elements in obtaining a well-made bell with a good sound are: the quality of the bronze alloy, weight, thickness, diameter, and height.

Right

A part of the Marinelli Bell Museum collection.

LA RAMERA, COPPER UTENSILS, MOLISE

"My utensils and pans are not only attractive but also very functional."

FRANCO GERBASI

LA RAMERA WAS started by Franco Gerbasi's grandfather, around 1900, initially to make copper pots and pans, but later he also began producing thin tin plate for tins in the emerging food processing industry. Franco has four brothers and a sister and initially did not want to work in the same field as his father, leaving the management to his siblings. They fared so badly that he gave up his formal education and, as a mark of respect to his father who died in 1983, took over the family business aided only by his wife.

Franco, his wife, and three workers are determined to re-invigorate copperware production in Agnone. A newly built large showroom on the edge of the town fully demonstrates his commitment. Luckily this investment coincides with a return to fashion of attractive copperware, especially in the kitchen where the more practical easy-care stainless steel pots and pans are visually less appealing. La Ramera produces not only traditional copperware for such uses as polenta making – large copper kettles were always preferred – but also the more practical *rame stagnato,* which has the more scratch resistant white copper on the inside while retaining beaten copper on the outside.

Above (left to right)
Skillet with leaf patterns on the base; mold with various fruits; decorative copper kettle, originally used directly over an open fire. Many of the motifs beaten onto the copperware are inspired by organic elements found locally in the surrounding countryside.

FRATELLI BARATO, ARTIGIANI DEL RAME, COPPERWORK, VENETO

Opposite (clockwise from top left)

Detail of a large beaten copper plate that shows the impressions of hundreds of fine and accurate blows to the sheet metal. Keeping the design concentric requires consummate skill. Their father, Ottorino Barato working on a copper kettle in the 1960s. Some of the tools, powdered paints, and small luminescent dishes produced by smelting and enameling copper plate. Classic copper bottles made by Ottorino from the early 1960s and an amphora-shaped vase inspired by the bottles made by Lorenzo in 1990.

NEAR THE NORTHERN town of Vicenza in the rich Veneto plain, two brothers are also following in their father's footsteps. The Barato brothers, Lorenzo and Luigi learned metalworking from their father, but before entering the family business, they both studied at the Institute of Art in Padua; Lorenzo specializing in sculpture and Luigi in metalwork. After their father's death in 1992, they developed different areas of activity, with Lorenzo dealing in the practical and Luigi, who also teaches metal and enamelwork in a local college, the design side of their business.

Their ancestors used to make agricultural equipment but around 100 years ago they turned to making household items: the usual pans, kettles, and milk churns as well as metal stoves. In the 1950s and 1960s, their business experienced a downturn with the arrival of central heating: old stoves used for decades were discarded completely, and, of course, new ones were no longer ordered. So the family had to adapt the business, or just close down. They initially turned to making decorative, rather than practical, items and soon found themselves making sculptures, copper tableaux, pictures, and plaques. Luigi also ventured into jewelry design. The richer north rediscovered old traditions much earlier than the poorer south and by the mid-1970s the brothers had noticed a slow growth in demand for kitchen items in copper including pans, oven hoods, and even stoves. Now the Barato brothers make a diverse range of copper-based products, using century-old tools (many of which belonged to their great grandfather), both decorative and functional, from huge church doors to copper coffee pots. One of their more extraordinary commissions was for a huge copper monument to commemorate those from nearby Padua who died in concentration camps in World War II. They also undertook the commission for two 10-foot (3-m) high doors for a local village church which took them two months to make.

About 15 years ago they introduced smelting and enameling of copper to some of their designs. Beautifully colored, luminescent effects are achieved by applying colored paints in powder form through little sieves to copper objects, which are then fired at a high temperature several times.

"Unlike our father, who made beautiful but basic utilitarian copperware, we are happy to venture into more artistic areas of work."

FRATELLI BARATO

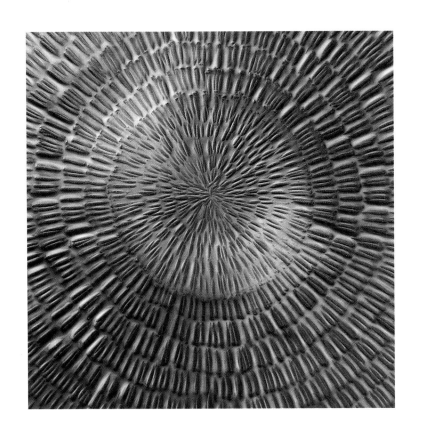

FLORA E FAUNA, DECORATIVE BRASS, TUSCANY

"We see our talents as natural and indigenous – how could we not be inspired by a city which exudes art and artisan skills in every form and on every corner, when, as children, our parents took us regularly around the streets and piazzas of the city, pointing out the wonderful churches and palazzi of the Renaissance?"

MASSIMO AND LUCA BENATTI

THE FLORENTINES, Massimo and Luca Benatti are two of four brothers from a family who worked in various crafts in their city. Indeed, they specifically chose a workshop and showroom in one of the narrow streets near the Pitti Palace where artisans have been working for centuries, furnishing and adorning the homes of wealthy Florentines.

Luca worked for a while in the fashion business in Milan and Florence before setting up Flora e Fauna in partnership with brother Massimo in 1994.

The idea behind their enterprise was simple: to make beautiful, elegant, and original objects for the home, that ordinary people could afford. For cost reasons alone, therefore, they decided to utilize brass rather than gold or silver as the basis for their products.

All their designs reflect organic motifs including flowers, leaves, and insects. Using thin stamped brass forms, made to their own specifications, they spot-weld various elements together to create intricate neo-Baroque pieces including trays, occasional tables, table and floor-standing lamps, picture frames, mirrors, bowls, dishes, planters, underplates, place setting holders, napkin rings, and table decorations.

Depending on how they feel about a design, and to some extent because there has always been a demand for "rich-looking" objects, they silver- or gold-plate the brass. These items go through various treatments before being varnished and dried in a special furnace. This last phase is important to prevent oxidation. Sometimes, however, the visible soldering spots are left to slowly oxidize and create natural shades and textures within the brass, leaving an organic feel to the finished product.

Above

All Flora e Fauna designs reflect organic motifs – flowers, leaves, insects, tree bark – which are here combined to form the borders of a picture frame.

Opposite
(clockwise from top left)

Spot-welding a stamped brass shell to form the first stage in a design.
Standard lampshade in silver plate using a combination of leaf motifs.
Close-up detail of a shell and starfish combination.
A completed galvanized silver plate with shell border, containing natural shells for contrast.

ARTIGIANI SANT'ELIGIO, SILVER-PLATED ITEMS, SARDINIA

"Because of our mining heritage we, in the south of Sardinia, have tended to make practical household items, whereas in the north, silver is traditionally used in jewelry and personal decoration."

PAOLO ESSENZIALE, CO-OP MEMBER

THE SMALL SARDINIAN artisan cooperative, Artigiani Sant'Eligio, in the historic southern medieval town of Iglésias (province of Cagliari), is an interesting pilot scheme. It is supported by ISOLA (*Istituto Sardo Organizazione Lavoro Artigiano* – the Sardinian Institute for the Organisation of Handicraft Work), where a group of silversmiths work is displayed and where exhibitions are sometimes held of other Sardinian crafts including ceramics, wood, lace, and crochet.

The recent closure of the mines from which metals such as zinc and lead had been worked since antiquity, has helped to stimulate other activities including the formation of the Sant'Eligio cooperative, some of whose members collaborated with Florentine silversmiths to learn about modern techniques based on traditional craftsmanship. Their newest work, such as the terracotta folklore figurines, is based on designs created with sheets of fine tin which are then silver-plated. The group find their inspiration in the Sardinian environment and local history, and, like the Benatti brothers in Florence, their designs feature flowers, fruit, and animals – as indeed do the street and family names in and around Iglésias.

Above (left)

The terracotta folklore figurines reflect modernized 19th-century styles by the use of silver-plate elements.

Above (right)

Detail of an underplate with graphic interpretations of local motifs.

VITTORIO MURA E FIGLI, METALWORK AND KNIVES, SARDINIA

"My sons are the fourth generation working in the tradition of making fine knives, but we are also committed to maintaining a skills base, even to the extent of producing the best possible implements for our ordinary customers."

MARIO MURA

Right

Two "shepherds' knives" both showing the simple elegance of this perfected cutting tool. Both have decorated cuffs; the upper blade is plain whereas the lower damask steel blade contains an elaborate inscribed motif engraved in gold.

ON THE WEST SIDE of Sardinia, in the small mountain town of Santu Lussúrgiu, province of Oristano, Mario Mura heads a small family business skilled in all areas of metalwork. With his sons, who are the fourth generation working in the business, they are particularly known for their tradition of making fine knives of such rare beauty that they are sought by collectors from all over the world.

Knives have always played an important role in a Sardinian society, dominated as it has been for centuries by hunters and shepherds. A man's knife was not only a functional tool, upon which his life often depended,

83

but also represented his status within the community. A traditional Sardinian knife also had to fulfill a number of functions equally well and so its tapering form evolved to cope with all situations – whether cutting wood, butchering an animal, or skinning a carcass. The Sardinian knife is a perfect example of form following function in one instrument.

As the shepherd became richer, he sought to enhance his knife in line with his status. The elaboration of both the handle and blade has created the collector's knife in which Mura e Figli are acknowledged masters. Their own collection, handed down through the generations, includes a 120-year-old hunting knife as well as fine examples of the family's own production.

A collector's knife might feature delicate carving on the horn handle, gold "cuffs" where the handle joins the blade and extremely fine blades of "damask" steel with gold inlay or engraving. To get a substantially straight handle out of a curved horn, a whole horn is often needed to make just one knife handle, which is the reason, together with the time and skill needed to produce the blade, that they are so costly. The "de luxe" versions of these knives can cost up to $4,300 (more than £3000).

Equally impressive, less expensive knives are made from cheaper, less interesting buffalo horn. Since the horn is longer and straighter, more handles can be made from one animal.

Top

Mario Mura forging a "damask" blade by folding red-hot special forged steel and hammering it flat repeatedly until the desired wave-form is achieved. It can take up to two days to produce this kind of blade.

Above

Mario cutting a horn. Knife handles were originally made from the curved horns of the *muflone*, a Sardinian mountain goat, but since these are now protected, horns are bought in. The natural flexibility of this material allows a particular "give" in the handle when the knife is used.

RAFAELE DI PRINZIO, WROUGHT-IRON METALWORK, ABRUZZO

Above (left)

Rafaele di Prinzio, a modern Vulcan forging a piece of iron in turns with his nephew. They often work in pairs and in rhythm to beat the hot metal more rapidly than they would be able to if working alone.

86

Above (right)

Close-up of the fine work that can be achieved in decorative wrought iron by highly skilled artisans.

"An idea must attract my attention in such a way that I can transform the fantasy into reality with a few simple movements. At night is when I can sit and work out how to form a new piece and, bit by bit, create it."

RAFAELE DI PRINZIO

THE DI PRINZIO FAMILY can trace their blacksmith's forge back to 1860 when Rafaele's great-grandfather started working as a *maniscalco* (blacksmith) in Guardiagrele, not far from Pescara in Abruzzo. From a modest beginning, the di Prinzios branched out into general ironwork, including knife making which required the use of tempered steel. Knives were socially so important in the countryside of Abruzzo that they were given as wedding presents until around 20 years ago. By day, Rafaele and his workers, nearly half of whom are family members, create head boards, iron gates, fire grates, candelabras, indeed anything for the home that can be made in wrought iron. "The work must be elegant, beautiful, and sincerely produced." says Rafaele.

At night, once all the workers have left, Rafaele returns to the forge and gets down to the serious business of wringing unusual forms out of the base material. The cold steel is brought to life in the forge's fire and, with the repetition of hammer on hot metal, the iron transforms itself into a cock, a windmill, or whatever else he might have in mind that day. Even when customers prefer finishes other than that of the "natural" wrought iron, di Prinzio insists that the coating does not increase the dimensions of the item more than is absolutely necessary. He demands that their products "... stay alive by showing the marks of the hammer blows which symbolically represent both the force contained within as well as the imprint of a true artisan at work."

JACKIE TUNE, DECORATIVE METAL, TUSCANY

"I very much enjoy the challenge of the larger creations – but I can only accept commissions from clients who will allow the organic process of creation and design to function my way."

JACKIE TUNE

JACKIE TUNE IS AN Italian by adoption: she was born in London, raised in Tunbridge Wells, and studied photography in Bournemouth. She has been living in Italy since 1990 when she came to join a friend who was giving photography workshops in Tuscany. It was after she rented an old house that she started playing around with old bits of rusty wire, plenty of which she found lying around the garden. Her first creations were little wire candlesticks. Friends liked them, so she made more and took samples to some of the shops in nearby Siena which sold them on her behalf.

A "wire boom" in the early 1990s made her original objects look rather ordinary, so Jackie started to experiment with thin metal, cut from sheets, to create interesting objects including candlesticks, bowls, and other decorative items. They were also well received, and she began

87

Previous page

A screen cut out from thin sheet steel and decorated using a combination of techniques. On the right is one of her standard wire candle holders.

Right

Jackie Tune's first creations were little wire candlesticks, which evolved from her experimentation with rusty wire found lying around the garden.

Opposite (above)

Patterns resulting from leaving sheet steel out in the rain. Depending on the amount of acid in the rain, the colors can be red, yellow, or orange.

Opposite (below)

Newer designs evolve from rust patterns formed by the rain.

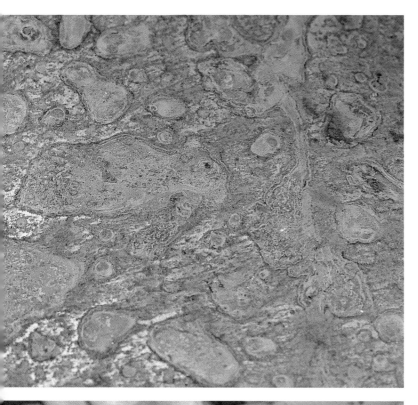

supplying shops in Siena, Florence, and Milan. From smaller items she moved on to bigger ones, adding standard lamps, candelabras, mirrors, tables, chairs, screens, garden furniture, and even fireplace frames, gates, and balconies with the inclusion of wrought-iron elements to her repertoire. Jackie calls it all "fairy furniture" in the sense that she often feels she is decorating a doll's house – albeit a human-sized one!

She met her companion, Claudio, in 1995. A builder by trade, from an agricultural family, born and raised in the province of Siena, Claudio works with Jackie on the heavier pieces, especially those that require beating and welding into shape. Before Claudio's arrival, she had to take pieces to local blacksmiths, and although she gave them precise instructions not to make things symmetrical and regular, they always did! Now she can oversee the shapes as they are being created and gets exactly what she wants.

89

An intriguing aspect of her work comes from the "naturally" textured look of many of her pieces. Different effects are achieved by the simple practice of leaving sheet steel out in the rain. The resulting patterns depend on the amount of acid in the rain and this affects whether the rusting turns red, yellow, or orange. Jackie says the "best acid rain" comes in the late summer and autumn.

Many of her newer designs develop from the rust patterns formed by the rain. Part of the pattern is removed with concentrated acid to produce the motifs she prefers, such as leaves and flowers, inspired by her rural surroundings deep in the heart of Tuscany. Once the desired effect is achieved, the applied acid is washed away before it etches into the metal sheet. Finally, the sheet is sealed with a transparent varnish to preserve the design. She says, "Sometimes I deliberately create the rust with acid, and I don't always just leave it to nature!"

Above (left)

A complex umbrella stand called
Veneri (Venuses).

Above (right)

The base of a streetlamp and an old
"soft" steel hammer.

Right

A wall-mounted fountain, without
the pipe.

FONDERIA CARNEVALE, ORNAMENTAL CAST IRON, LAZIO

"Perhaps we were lucky because we were the last to arrive in this sector, and my father did not try to compete with the bigger factories. He wanted to make things to order and preferred the ornamental to the strictly practical."

MARIO CARNEVALE

THE FONDERIA CARNEVALE was founded by Luigi Carnevale in 1949 in a zone north of Rome's *termini* (central railway station), which had been badly bombed during World War II. The streets around *Campo Verano*, Rome's huge main cemetery near to one of ancient Rome's gates, have been a center of artisan activity for centuries. Artisans worked near each other, specializing in one craft while often depending on another skill for some element in their output. As Mario, Luigi's son explains: "An ornamental foundry such as ours needs, for example, someone to model our designs in wood so that we can create the stamp into which we pour the molten iron."

The phases of the Fonderia's work faithfully adhere to 19th-century methodology when the reproduction of cast-iron elements in quantity became established. "Apart from the introduction of more modern power tools to help in the assembly process, the only real difference between then and now is that we use aluminum molds to make our facsimile products. The initial designs are still made in wood first."

The Fonderia Carnevale workshop in via Sabelli may no longer produce the molten product but they still assemble to order whatever their clients, mainly architects and designers, are able to invent. Fonderia Carnevale make a range of tables, chairs, fountains, fireplace frames, balustrades, spiral staircases, and traditional 19th-century postboxes. Mario Carnevale is always searching for new ideas, scouring local flea-markets, such as the famous Porta Portese in Rome, for forms or shapes that may become useful in some future design.

Above

An ornamental frieze that can be used in a fireplace, on a gate, or along a balustrade.

textiles ~ weaving stories

In Italy, textiles range across the whole spectrum – from the luxurious silks, damasks, and brocades famous since the Renaissance, to the rural and peasant weaving traditions in Sardinia and other areas of the south. Even in the arena of mass production, Italy is still one of the dominant world forces in textiles through its fashion industry.

At both ends of the spectrum, these textiles have a story to tell – of the generations of women who make and have made them, and the recipients and users of them. In the south and Sardinia, the majority of weaving was done for purely practical reasons, to make everyday objects for the home and working wardrobe, and decorative items were made either for a daughter's dowry or the church, where they were accorded pride of place for display rather than use. A world away in Venice and Florence, the hand loom weavers are still producing lavish cloths for the combination of practicality and decoration in the homes of the rich and famous, as well as the world's most celebrated interior design and fashion houses.

Textile "art" is a sensuous medium – whether it be the chunky robustness of Sardinian wool rugs, the luxurious pile of Venetian velvets, or the luminous sheen of Florentine silks. The feel is an integral part of their beauty – more so than all the other crafts of stone, ceramics, metal, and glass whose texture is more important for the way it looks rather than the way it feels. Also, textiles are soft and malleable and, therefore, susceptible to both light and movement as enhancement of their intrinsic qualities.

opposite: Sumptuous hand-printed velvet in traditional patterns from the Fondaco di Venezia by Lake Garda.

Above

At Villanova the weavers introduce modifications to the traditional *biscotto* (plain biscuit color) rugs by weaving in naturally dyed colors.

COOP. TESSITRICI, TRADITIONAL WOVEN ITEMS, SARDINIA

"We work at the looms for love, and because we are committed to preserving our Sardinian heritage and traditions. Without the backing of ISOLA and husbands and families to support us, we could not survive by weaving alone."

SIGNORA PES

THE GEOGRAPHIC NATURE of Sardinia's isolation in the Mediterranean has always necessitated self-sufficiency and making the best use of natural resources. Unlike the artisans of the cities and trade centers of mainland Italy, Sardinians made craft objects for their own use: basic items for everyday life and decorative items usually only for local religious and festive occasions. Inspiration for their designs comes from the local flora and fauna and the myths and legends of their heritage; the lack of cultural exchange gives Sardininan crafts a very genuine identity.

Only recently has growing demand for Sardinian handwoven carpets and household linens and furnishings spread to the rest of Italy and overseas, largely because of improved transportation and communication and the growth of tourism. The materials used are local wool, flax, cotton, and linen, and the dyes are often still obtained from infusions of leaves, flowers, bark, and roots, from colored earth and ochre, and even the seaweed around Sardinia's shores.

In the northwestern village of Villanova Monteleone, a small co-operative of women is provided with a workspace by ISOLA – the Sardinian Institute for the Organization of Handicraft Work, a public body created in 1957 to support and promote the technical, artistic, and commercial development of Sardinian handicrafts. Artigianato di Villanova is housed in a large open schoolroom-type building furnished with privately owned looms so that the local women have the opportunity to work in company, rather than solo at home. They organize their own working hours, often bringing their small children with them. The majority have learned their craft at locally run courses, also funded by ISOLA. Their output ranges from small woven rugs to large carpets and wall hangings.

Left

The textile products from each village tell a story, because all are differentiated by their patterns, colors, and subject matter. In Sant'Antioco, the weavers generally employ the flat-weave technique and use traditional colors; in Villanova Monteleone, the designs are largely geometrical and more plainly colored. The weaving techniques employed all over Sardinia are limited to a relatively simple inventory and are carried out mostly on horizontal looms.

Right

The weft yarn is wound around a shuttle that passes through the warp, unwinding the yarn as it moves.

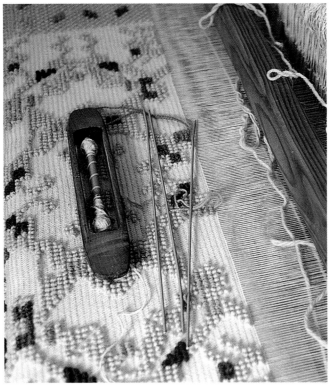

Above

Pibione or "grape" weaving is the most typical Sardinian method and is achieved by twisting the warp yarn on a wire placed horizontally on the loom. The "grapes" are stopped by the passing of the weft and fixed in place with one or more beats of the sley (the baton brought down the horizontal loom to compact the weave). The wire is then withdrawn to give a relief effect – like grapes.

ANTICA STAMPERIA MARCHI SANT'ARCANGELO, HAND-PRINTED LINENS, EMILIA ROMAGNA

"In our shop, which dates from 1600, we continue to stamp fabrics with the same colors and use the same designs and techniques as those of the traditional folk art of the Romagna."

ALFONSO MARCHI

HANDBLOCK PRINTING on hemp and linen is one of the oldest ways of decorating fabric and was first developed in the Far East. The old family firm of Antica Stamperia Marchi Sant'Arcangelo in Emilia, decorate old and new cotton and hemp fabrics to produce robust cloths for household use. The tablecloths, aprons, kitchen towels, and place mats are distinguished by local rural designs applied with hard pearwood blocks in natural dyes – a range of earthy, rusty reds, and browns being particularly distinctive of the Romagna region.

IL FONDACO DI VENEZIA, HAND-PRINTED VELVET, LOMBARDY

"I have no precise recipe or instruction booklet for my work, just years of experience and a practiced eye."

ROSANNA SCHIAVON

ANOTHER FORM OF stamping with hand-engraved blocks is practiced by the Schiavon family. Claudio and his wife Rosanna used to make furniture in Venice but in 1988 decided to move into the soft furnishing side of the business – and also away from the stifling and crowded center of Venice. They set up the Fondaco di Venezia on the shores of Lake Garda, continuing their inborn attraction to living by the water, as well as their intrinsic love of the sumptuous fabrics associated with Venice. Here, in their three-story

Above (left)

Claudio Schiavon designs and carves the woodblocks: his sources are art and history books, old fabric fragments and samples as well as the classic Venetian patterns which the majority of their clients want.

Above (right)

For the paint, an acrylic base is mixed with inks and metallic powders, overlaid on the block with a smattering of gold to lend sheen and a luxurious antique effect.

workshop, rooms are stacked with bolts of richly colored silks and velvets, shelves are stuffed with softly patterned cushions and quilted coverlets, and walls and windows are adorned with plush drapery catching light and shadow in its folds. The muted, almost faded colors of the fabrics and printing lend an antique quality, although they are new.

Their skill is the decoration of the handwoven fabrics – silks and silk velvet – which are bought from Canavese near Turin and dyed to colors specified by them by a company based near Pordenone in the Friuli, northern Italy. Claudio designs and carves the woodblocks, Rosanna mixes the paints, using an acrylic base mixed with inks and metallic powders. She keeps no record of her measurements and quantities. If a client wants a replica of or an addition to something they already have, they have to send a sample for her to match.

Having laid out the material and measured the positioning of the pattern, she and a helper sponge the paint onto the lino wood-block. For a large pattern, two people are needed as the paint dries quickly (for the same reason, the paints are mixed in small quantities for each job). On top of the color, they add a smattering of gold to lend a sheen to the design; then the block is carefully turned over onto the fabric and struck with a wooden mallet before being expertly lifted off. The woodblocks are then immediately washed with water, before repeating the process for the next part of the pattern. The painted fabrics are made into coverlets, curtains, table runners, cushions, bed-heads, and wallcoverings with decorative tassles, braids, and trims added to complete the luxurious finish.

99

Above (top)

Rosanna and her assistant are both deft and speedy in sponging on the paint as it dries very quickly.

Above (below)

The painted blocks are placed on the velvet and struck with a wooden mallet, gradually building up an intricate pattern on the material.

ANTICO SETIFICIO FIORENTINO, LUXURY SILK, TUSCANY

"The luxury of going slowly, which gives these fabrics an inimitable quality, requires, besides manual skill and a profound knowledge of the job, a passion for doing it."

SABINE PRETSCH

IN THE HEART of the artisan district of Florence, the spirit and work of the Renaissance lives on: the world-famous Antico Setificio Fiorentino, brought to prominence again after World War II by Emilio and Alessandro Pucci di Barsento, is now presided over by Sabine Pretsch, who is a passionate custodian of Florentine silk traditions and the art of hand weaving, although she herself has no background in weaving. She has lived and worked in Italy for almost 30 years: following an initial period in Rome studying theater and set design, she drifted into journalism but still pursued her deep love for Italian arts and culture – a love that eventually drew her to the Renaissance city of Florence. While on assignment for a radio program on the Antico Setificio Fiorentino, Sabine found herself in the middle of a company crisis and a threat of closure when the managing director suddenly resigned. With no other suitable candidate, and recognizing Sabine's appreciation of the company's worth, Signore Pucci offered her the job, and she spontaneously accepted.

As in weaving itself, Sabine truly believes in the "luxury of going slowly," and has built the customer base and profile of the company steadily and qualitatively, capitalizing on the knowledge and dedication of her workers to meet demand for traditional silks, as well as to create new fabrics and techniques. Such is the skill these artisans harbor (design, dyeing, and weaving are all done on the premises), they can satisfy any request and are unfazed by demands such as, "I want a silk in red like the color of dried tomatoes – not the tomatoes of Naples but the ones from Palermo" (a world famous interior designer!). In the last five years, they have noticed a growing trend for greater exclusivity and personalization of goods – a demand that is, of course, satisfied by handmade goods.

The looms are worked exclusively by women and the aura of unity and comradeship is palpable. The only man there is the mechanic who has a very special skill – because when a part breaks on an 18th-century wooden loom the only way to fix it is to make the replacement part. Most of their looms date back to the end of the 18th century; the oldest is a small 16th-century Renaissance loom on which they weave fringes and braids. The silks produced at Antico Setificio are unique because each loom is individually designed: one example is a traditional Tuscan rustic weave that appears to be solid red but is in fact made up of between 17 and 54 different shades of thread. Sabine's boast is that they can custom produce any fabric design suitable for a Renaissance palace or a rustic farmhouse.

Above (left)

Silk ribbons and braids are woven on a small 17th-century hand loom in patterns from the archives. Here they are interlaced to make a cushion cover.

Above (right)

Setificio's most experienced weaver working on a gold damask of a design that dates back to the Renaissance. Using 48,000 threads, a full day's work will progress the fabric 8 inches (20 cm). This loom is the one on which the apprentices take their graduation "exam" after their five-year training period.

Left

The flood in Florence in 1966 destroyed many of Antico Setificio's archives. This design from the 18th century survived and is still being produced today.

TESSITURA BEVILACQUA, SILK AND VELVET, VENICE

"Our weavers know when the waters are going to rise in St Mark's Square several hours beforehand because of the way the wooden looms begin to squeak and the way they react to the changing humidity."

RODOLFO BEVILACQUA

TESSITURA BEVILACQUA is a family-run business that has been operating in Venice for 200 years. The brothers Rodolfo and Alberto are the proud inheritors of one of the city's most celebrated silk-weaving concerns, only working to commission and having produced pieces for the White House, the Vatican, and royal palaces in Sweden, Belgium and Kuwait. More than 3,500 unique designs – from 13th century to Art Deco – are stored in sample books stacked from floor to ceiling in an ancient wooden shelving system in the entrance to the main factory room.

Opposite

Nowadays these sumptuous damasks and brocades are generally used for furnishings, although they were originally created for religious vestments or the garments of the aristocracy and wealthy Venetian merchants.

103

The workshop is situated in Santa Croce, just behind the Grand Canal, and all their weaving is done on 15 antique looms, which were rescued by a Bevilacqua antecedent from the School of Silk in Misericordia when it was closed down by Napoleon. It can take between 15 days and three months just to prepare the loom, before weaving begins.

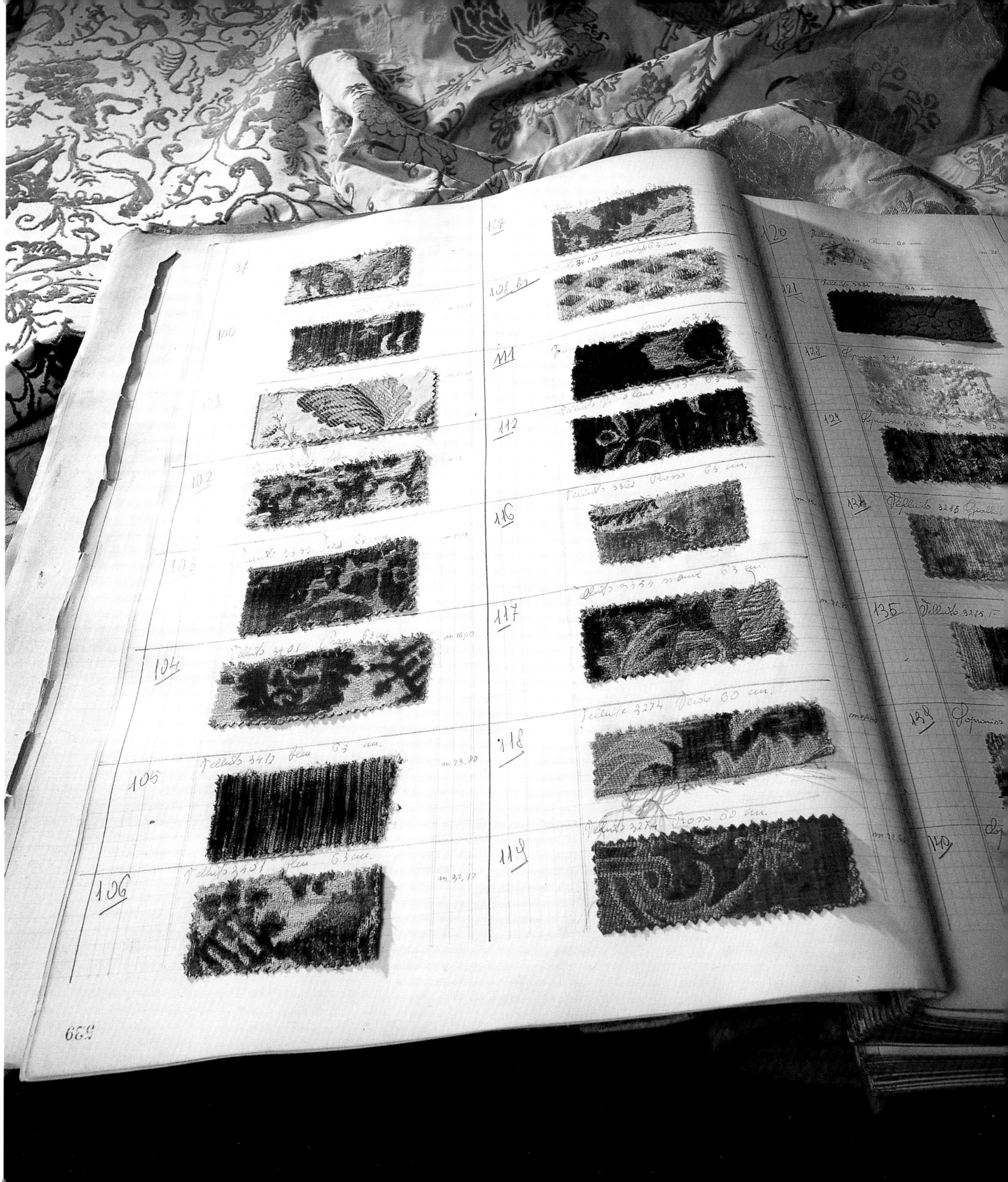

Left

This red on gold design is an extremely expensive brocade because it does not repeat and decorates the famous Venetian Church of La Salute on the Grand Canal at festival time.

Opposite

More than 3,500 unique designs, in the form of punched cards, or "cartoons" as they call them, are stored in the workshop. This sample book of swatches records some of the materials produced by the Bevilacqua factory.

Left

This *sopra-rizzo* design fabric is set up with 800 bobbins, 6,300 threads and four colors, in addition to the background. The purchase and dyeing of the raw silk for the Bevilacqua house is all carried out in Venice.

ARTEVIVA, WOVEN TEXTILES AND TAPESTRY, FRIULI

"I love to mix textures and create unusual fabrics and effects by weaving together wool, silk, nylon, and metal filaments."

LIVIANA DI GIUSTO

IN FRIULI, in the northeastern corner of Italy, Liviana di Giusto, who trained in textiles at the local Institute of Art, brings a very contemporary accent to the ancient art of hand weaving. Arteviva, located in the town center of Udine, is the small shop front and workspace into which she has squeezed her two wooden looms and a third vertical loom specifically for weaving wall hangings.

With a keen eye for color and a deep appreciation of texture, she loves to experiment with a diverse range of natural and synthetic fibers including wool, silk, metal filament, nylon, polyester, and plastic. She creates tactile works of art: wall hangings that are an assemblage of textures and layers of thread – coarse and tweedy and mohair-soft – representative of her unique interpretation of the natural elements of fire, earth, air, and water which are her inspirational themes. With the spark of an idea, these designs usually develop on the loom as she works. In this area she is happiest, but the need to make a living means servicing a more everyday market with smaller items such as scarves, wraps, hats, and cushions. Demand for larger artifacts is minimal because paintings or sculpture, rather than textiles, tend to be preferred for commissioned works.

106

Far left (top)

Ideas and color combination develop on the loom, here with a loose woolen weave which sees the fabric grow quite quickly.

Far left (below)

Liviana's looms are all modern and were made to her specifications by her carpenter father-in-law.

Left

A designated theme of "crucifixion" for a craft show was the inspiration for this wall hanging by Liviana di Giusto, whose use of fiber and texture is a powerful medium of artistic expression.

107

glass ~ trial by fire

Artistry, imagination, and technical skills have imbued glass with a seemingly limitless variety of form, color, texture, and luminosity for thousands of years, and Venice is undeniably one of the most influential and important centers devoted to its production. The Venetian glass furnaces were moved to the island of Murano in 1291, officially to reduce the risk of fire in the city, but also to keep trade secrets away from prying eyes. By the mid-15th century, Venice had become the predominant glassmaking center in the world – both in technical and artistic terms – renowned for its glass-blowing skills (*vetro soffiato*) in particular. Murano glass came to dominate the European industry and flourished unchallenged from 1400-1700.

The delicacy and transparency of glass make it a precious and treasured object in any form it takes, yet its actual make-up is of easily accessible, basic ingredients – a fusion of fire, sand, soda, and lime. Its technical description as a "rigid liquid" gives some idea of the special qualities it possesses, as well as the skills required to control and mold such a volatile substance. Unlike the raw materials of fiber, wood, stone, and clay, glass has to be made and is not as amenable to slow and considered crafting. With no margin for error, glass blowers have for centuries explored the aesthetic, functional and cultural vocabulary of the medium, literally in the heat of the furnace with split-second intuition and control.

Involving more than just the hands, glass blowing is literally a product of breathing, a crucial part of a maestro's experience and skill.

opposite: Pier Lorenzo Salvoni's butterfly in glass and copper wire, resting on one of his elaborate glass candelabras.

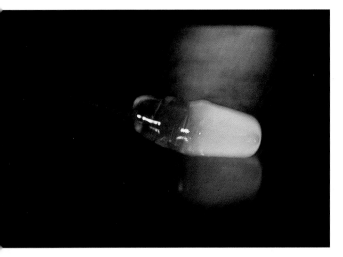

BAROVIER & TOSO, CLASSIC GLASS, VENETO

"It is often the simplest, most unadorned objects that are the most difficult to produce. An idea on paper can fall apart when put to trial by fire."

ANGELO BAROVIER

ONE OF THE FINEST examples of family commitment to maintaining a handmade craft through generations, Barovier is recognized as the world's oldest glassmaking family in the *Guinness Book of Records*. Practically as old as Murano itself, the Barovier name has been associated with glassmaking since 1295.

By the 16th century there were three Barovier brothers operating their own individual furnaces in Murano, each with their own trademark: an angel, a star, and a bell – all of which later came together to form the family crest still in use today. While the fortunes of the Baroviers reflected those of the island of Murano (indeed, Venice), which went into decline at the end of the 1700s, they saw a revival under the leadership of Giovanni Barovier and his three nephews, Benvenuto, Benedetto and Giuseppe in the latter half of the 19th century. Ercole Barovier (son of Benvenuto) brought fame to the family again in the early 20th century, experimenting with color and ground-breaking new production techniques. As one of the pioneers of the *Forme Nuove* period in 1950s glass design, Ercole is credited with introducing the use of thick, heavy glass and a "mosaic effect" created with heat-induced coloring without fusion. This classic design is still in production and in demand today. Ercole was a true maestro – both at the kiln and at the drawing board. This is not always the case: many designers have brought their design skills to the furnace and then had to learn the temperament of their raw material and how it is handled by the maestro, because only he could execute the final product.

Nowadays Barovier & Toso is still producing traditional forms – from fabulously ornate and intricate chandeliers to the vases and bowls of heavy air bubble glass and colored mosaic that were introduced in the 1950s.

Opposite (top to bottom)

The creation of blown glass is a carefully choreographed team process. Considerable physical demands are placed on the glass blower whose timing and intuitive appraisal of how form is emerging is as crucial as his ability to interpret design.

Right

An early Ercole Barovier design first created in the 1930s, using thick walled glass to give the idea of bulk as well as presenting the quality of the glass in a more three-dimensional and visible way. During the same period, he also introduced another innovation called *colorazione a caldo senza fusione* (coloring while hot without fusion occurring). At the time, the company traded as Vetreria Artistica Barovier, becoming Barovier & Toso in 1937, and Ercole was unusual in that he combined the function of glass technician and designer. In previous centuries, glass produced on Murano was essentially a collaboration between owners and their employees.

Above (left)

The deft use of pincers and scissors form handles, stems, and decorative effects, while subtle body movements and sudden twists can provide the individual signature on a piece.

Above (right)

This fabulously ornate and intricate floral chandelier, exploding with color and light, is a characteristic example of traditional 17th-century Venetian opulence (Barovier & Toso).

Left

An early Ercole Barovier design first created in the 1930s, using thick walled glass with the introduction of air bubbles. This type of vase became a hallmark design for the firm and has been popular for many decades.

FRATELLI BARBINI, VENETIAN MIRRORS, VENICE

Below (top to bottom)

Various applications of cerium oxide, silver nitrate, soda, and ammonia are punctuated with copious rinses in distilled water to produce pure unmarked mirrors that are then assembled and mounted using fine wire and pins onto wooden bases.

ANOTHER CENTURIES-OLD Murano tradition is the making of mirrors – a form which, along with chandeliers, demonstrates the heights of fantasy and opulence achieved in luxury glass production. Having perfected the amalgams of tin and mercury required to achieve supreme reflectivity and permanence, the Venetians held an unassailable market position in mirror-making from the early 16th to the late 17th century. Venetian mirrors, extravagant masterpieces made up of hundreds of varied glass and mirrored fragments and florets are still made today – not just for palaces and staterooms, but for shop fronts and private apartments worldwide.

The current generation Barbini family of seven children traces its roots back to the mid-1500s in Murano, although the current business was founded by an uncle, an ordinary glassworker who taught himself engraving. After World War I, with the help of the whole family, he began his own business in engraving and mirrors. Several of the brothers are now practicing their skills in the United States, while two more uphold the mirror-making traditions and oversee new generations in Murano. They can execute any design a client requires, but the majority of their output is rooted in the ostentatious designs of the 17th and 18th centuries favored by a contemporary clientele.

The Barbini firm employs a team of 10, each an expert in one area – designing, silvering, etching, cutting, assembling, mounting. The engravers etch freehand designs onto the glass shapes using stone wheels under constantly dripping water and then pass the glass over, to go through the silvering process of turning it into a mirror.

"We are always open to new ideas and happy to listen to what our clients want, because we have the know-how to execute practically anything. We are just doing what we know and grew up knowing."

GUIDO BARBINI

113

114

Right

One of the more
fabulous mirrors made
by Barbini. The
brilliantly attired
Moorish or Saracen
figurine was often
used in various guises
during the 16th and
17th centuries in
Venetian villas. Here
it has been turned
into a standard lamp
with an ornate
candelabra.

Above

The more ornate frames around the "looking glass plate," are made up of hundreds of glass plates, rods, beads, and ornamentation joined together with tiny glass-headed pins.

Left

Two examples of the diversity of effect that can be achieved by combining the many different elements in mirror design.

BOTTEGA VARISCO, ENGRAVED GLASS, VENETO

"I owe everything to my father: he taught me the technique, but more than that, he transmitted to me his energy and moral strength."

ITALO VARISCO

IF IN THE LATE 18TH CENTURY the dominance of Venice in world glass declined because of new types of glass being invented elsewhere, then the Varisco family business of glass engravers has gone some way to correcting the balance in the 20th century, having pursued a glassmaking technique actually made famous by the British. Italo Varisco has an array of international prizes and awards including the prestigious titles of *Cavaliere della Repubblica* and *Commendatore* conferred by the President of the Italian Republic, and the San Liberale Prize awarded by the Mayor of Treviso to those who have made the town famous through their art and craftsmanship.

Italo was born in Murano in 1940 into a glassmaking family which moved to the mainland town of Treviso when he was 14. From a small showroom and workshop attached to their house in an ordinary residential street, the Varisco family serve a worldwide clientele. Customers walk straight into the workshop to the metallic sounds of glass etching and cutting on stone wheels. Along the sides of the room the engravers (Italo, his son Marco, and three workers) are busy at their wheels, engrossed in their precision work, while behind a small central display the packing and shipping is done by Italo's daughter, Christina, and colleagues. Italo's wife, Bruna, takes general charge of the staff, the office, and the first-floor gallery where her husband's and son's masterpieces are on display, along with some historical examples made by Marco Varisco senior, Italo's father and founder of the family glass-engraving business (including a set of cut-glass vases commissioned for Mussolini in 1938).

Marco Varisco senior began experimenting with engraving on glass in the 1930s and subsequently won gold medals for his achievements from the Murano School of experimental glassmaking in 1938 and 1941. Glass engraving is another skill originally perfected by the Italians: Vicenzo di Angelo dall Gall is credited with introducing this decorative technique in the mid-16th century using the point of a diamond, hand-held like a pen. This technique was perfectly suited to the fine crystal glass and copied everywhere, but was challenged by the invention of lead crystal which could be cut and engraved on an abrasive wheel. In the second half of

116

Right (top)

In the Varisco workshop there are over 1,000 different abrasive and engraving wheels of various dimensions and finishes, some of them over 100 years old.

Right (bottom)

The glass cutter has to work "backwards" on the side furthest from him, through a trickling stream of water and the side of the glass nearest to him.

Far right

This huge vase, approximately 30 in. (76 cm) tall, shows off brilliantly the deep translucence and light-refracting qualities of lead glass.

Top (left and right)

After engraving or facet-cutting large sections of glass, it is often necessary to polish the opaque elements so that the surface becomes sufficiently transparent to enable the engraver to move on to the next section. A final polishing takes place after the piece is completed and various techniques are used to achieve different finishes.

the 18th century, the division in glassmaking became most pronounced: some worked with glass as a vitreous substance to be molded and formed while hot; others cut and engraved glass in its solid state, much the same as stone or crystal. The former method was at the heart of the Venetian glassmaking supremacy but, with the emergence of lead glass, a new age of cut glass arrived with more architecturally inspired designs and shapes.

Watching Italo Varisco at the grinding wheel is like watching a sculptor at work. He takes huge unadorned vases weighing as much as 22 pounds (10 kg) and, with a combination of incredible strength, precision, control, and freehand artistry, produces a finished object weighing 13 pounds (6 kg). Even more remarkable is that the glass cutter has to work "backwards" on the side furthest from him, through a trickling stream of water and the side of the glass nearest to him, and the only design he has to follow is the one in his mind's eye, for no mark is made on the glass beforehand. This three-dimensional facet-cutting process, followed by polishing with lead oxide, cork, and pumice, shows off brilliantly the deep translucence and light-refracting qualities of lead glass.

Engraving on smaller items is like watching an artist sketching, except instead of moving his brush over the canvas, he is moving his "canvas" over the "brush" – in this instance a vertical stone cutting wheel. In the Varisco workshop there are more than 1,000 different wheels, some of which, together with various of the stone grinding machines, originate from Marco senior's business and are more than 100 years old.

Marco junior is passionately following in the family tradition, but has evidently developed his own style of expression and respect for the beauty and versatility of glass. He began working with glass at the age of eight and enjoys putting his own interpretation on traditional styles.

PIER LORENZO SALVONI, DECORATIVE GLASS, LOMBARDY

BREATHING NEW LIFE into glass design is Pier Lorenzo Salvoni. He creates utterly decadent glass objects resembling giant jewels in the unlikely setting of Pompiano, a small agricultural town deep in the industrialized northern region of Lombardy, where fog and the odor of the surrounding farms seem all-pervasive. In the attic rooms of an unassuming family house where he lives with his parents, Pier Lorenzo works alone in an environment that is calm and orderly and where color-graded rows of containers full of *cotisso*, pieces of broken glass left over from other processes can be seen. Using simple glass artifacts – bowls, plates, and vases – as his base, he cuts and carves these translucent fragments, and with golden rings,

"I love working on my own. After the chaos and hubbub of Venice where I trained, I really appreciate the peace and solitude of an orderly environment in which to create my pieces."

PIER LORENZO SALVONI

119

Left

Cotisso, chunks and splinters of broken glass swept up from the Murano glass factory floors and sold off by the kilo, are the raw materials that inspire Pier Lorenzo Salvoni, with their shapes and colors to create his "hanging jewels" and glass insects.

Left

Pier Lorenzo cuts and carves the translucent *cotisso* and attaches the resulting shapes to simple glass pieces, such as bowls or vases, using golden rings, chains, wires, and richly colored Florentine silk tassels, to create utterly decadent decorations for the home.

Opposite

An elaborate candelabra such as this is a natural progression from the jewelry Pier Lorenzo used to make when he first started working with glass.

chains, and wires and richly colored Florentine silk tassels, creates hanging pendants and fabulously inventive candelabra.

Pier Lorenzo went to Venice at the age of 18 to study architecture. He was not influenced by family tradition – they have always worked in wine and farming. Following his university education, he began working in glass, mostly making jewelry and other small items such as candlesticks. The opulence and luxury of Venice, both today and in bygone times, inspire his work and his attitude. He refuses to follow the market and creates only according to his own desires and motivations. This attitude comes from integrity (rather than pomposity) and imme-diate success: his works are regularly exhibited and sold in galleries in Rome, Florence, and Milan.

TIFFANY'S STUDIO, TIFFANY-STYLE AND FUSED GLASS, SICILY

FROM ONE UNLIKELY SETTING for glass art, to another – Syracuse in the southernmost corner of Sicily. Only unlikely in this instance, because glass artist Giuseppe Santoro also comes from a family background totally unconnected with the craft – he originally studied law at Sicily's Catania University. His interest, primarily in Tiffany glass, stems from his time studying art and design in Karlsruhe, Germany, then traveling in Europe and dabbling and experimenting with glass craft. His natural affinity and appreciation of the medium are evident in his eclectic range of designs, from typical turn-of-the-century Tiffany replicas, to fused glass wall lamps in the shape of mask heads and in his experiments with ceraics.

Giuseppe is skilled in all areas of glass panel work, or "Tiffany art," but his workers are generally employed for specific stages in the process.

"Despite having employees to help on various parts of the process, I am essentially still a one man show, because I am the only one who knows, and can execute, every part of the process in making these glass artefacts."

GIUSEPPE SANTORO

122

Opposite (clockwise from top left)

The Tiffany style is defined by vibrant and bold designs. Here, Giuseppe demonstrates two departures from the original inspiration of Tiffany – a portrait, a stratified lampshade, as well as a pure Tiffany panel, and an elegant fused glass invention which shows what can be achieved with skill and care.

Right

Giuseppe orders in glass from Oregon in the United States because it is important to ensure a uniform thickness .12 inches (approx 3 mm) for this craft and also because this particular manufacturer provides a huge color range in both translucent and opaque varieties. Classic Tiffany often includes floral motifs whether used pictorially or forming the physical shape of lamps and other artifacts such as vases.

ARTEVETRO, FUSED AND LEADED GLASS, SARDINIA

"I wanted to revitalize the contemporary language of glass by bringing out the various chromatic effects of the material through its transparent qualities."

LUCIANO GOLA

ON ANOTHER ITALIAN island and dedicated to the same glass art is Luciano Gola. Artevetro di Luciano Gola, based in Alghero in the northern part of Sardinia, specializes in handmade individual fused glass pieces as well as doors and screens utilizing the "Tiffany" procedures.

Luciano was born in the nearby provincial capital, Sassari. Before he specialized in glass design, he studied at the State Art Institute where applied arts such as jewelry, coral work, and *pietra dura* were taught. To begin with, he worked in the field of oil painting restoration and conservation, but discovered a growing fascination for glass and its artistic possibilities which, he noticed, required similar techniques. In time, and partly because of a deepening interest in the archaic, he began to work with glass-fusion (*vetrofusione*), which he realized was the oldest glassmaking technique but had been virtually forgotten in the modern world. He works alone in his neat and well-organized modern studio, on the outskirts of Alghero, designing and making by hand refined glass objects of considerable style based on original and reinterpreted ideas.

Left

Luciano Gola is especially interested in colors and techniques that either echo the past or translate well into modern design. He is constantly pushing the boundaries of fused glass through experimentation and research.

Right (top)

One special technique allows the introduction of bubbles into the fused glass process.

Right (below)

Shapes are produced using special ceramic molds.

applied arts ~ practical & ephemeral

The "decorative arts" contained in this section of the book are representative of a wide variety of handmade pursuits, most of which have deep roots in the traditions of the Italian countryside. The exception is the inclusion of painted eggs, particular to the northern, Germanic parts of Italy, indicating Central European influences.

Two of the other chosen handmade products are very specific to their regions: Sicilian painted carts and the terracotta whistles of Rutigliano, near Bari, in the south. Strangely, it is the whistles that have the longest as well as the most mysterious history, even though the modern versions have been revived by the national annual festival and competition held in Rutigliano for more than 10 years. The narrative elements of Sicilian cart painting are mostly 19th century but, even here, the roots of the coloring and figurative expression are synthesized versions of Sicilian-Arabic folklore.

Of the remaining two, the art of the "fresco" has been synonymous with Italian artists for more than two millennia. It is also generally accepted that the working of leather, in all of its many guises, is also one of the supreme Italian skills and Italian fashion output dominates the world leather goods market.

opposite: Painted goose eggs, part of a series of exquisite miniatures of spring flowers and butterfly wing motifs by Elsa Kondrak from Bolzano.

PATRIZIA PINNA, LEATHERWORK, TUSCANY

*"I am happy to be working on my own, for myself, and to be creating
with my hands. I enjoy the luxury of a slower-paced lifestyle."*

PATRIZIA PINNA

FINE LEATHER HAS always been synonymous with style and
fashion – never more so than today. Now, as in the Renaissance,
Italian artisans are at the forefront of fine creative leather work.

Patrizia Pinna, born and raised in Rome, decided to escape
the hustle and bustle of the city early on in her adult life. She
came to Sorrano, the small Tuscan hill town of her
grandmother's birth, and began an apprenticeship with an
elderly local artisan working in leather. Here she learned not
only the basic skills of dyeing, cutting, and sewing, but also a
keen appreciation of leather and how to nurture its best
qualities. Perched high above a gorge, Sorrano is a picturesque
historical town and frequented by foreign tourists – so Patrizia
has been able to make a living from her skills.

For 18 years she made traditional bags, wallets, and briefcases
in the burnished, red-brown leather, typical of the Florentine
style, but recently she has begun creating her own personal
designs. She mostly works directly onto dampened leather,
improvising as she goes along, rather than starting with drawings.
Normally, embossing is done by "branding" the leather with a
hot template, whereas Patrizia's method takes much longer but
allows her creativity to evolve as she progresses.

Left (above)

To produce the
embossing, the leather
is dampened and
stretched over a tem-
plate then, using a
range of implements
made of bone and
Bakelite™, Patrizia
pushes into the
leather to create the
indentations and
grooves.

129

Left (below)

Once the design has
been embossed,
Patrizia paints directly
onto the leather to
accentuate the
patterns.

Opposite

Patrizia Pinna's
stunning handbags
have a distinctive style
– unusual shapes with
striking abstract
designs and subtle
variations of texture
achieved by stitching
and embossing.

130

The leather she uses is organic, and apart from the black leather, which is bought from a reliable supplier, it is dyed locally with natural vegetable dyes. These dyes from local flora produce the deep burnished red-brown color that has made Tuscan leather recognizable throughout the world. She abhors anything artificial and only works with leather where no chemicals are used in the tanning, a process that generally takes from 20 days to a month, as opposed to two or three days if done industrially. Because the leather is organic, it retains more of its natural fats, and, therefore, lasts longer and actually improves with age. The ceramic

Above

Two leather-framed mirrors, one of which is Klimt inspired, reflect new directions in which Patrizia Pinna is taking her craft.

clasps and buttons are made by a fellow artisan, Maura Funghi, who has a workshop a few streets away.

In 1998 Patrizia first exhibited her collection at local fairs in and around Tuscany. Her work is taking on new shapes and sizes – as well as the handbags, she has added book covers, large-scale picture and mirror frames, and decorative plaques to her collection. She has had no formal art training, but one cannot help but wonder at her modesty and why it took so long for her imagination to flourish. Like a fine wine, the longer and less hurried the maturing process, the more superior the product.

ALESSANDRO MARCHETTI, LEATHERWORK, TUSCANY

"Without the inspiration and encouragement I received from two local ladies who wished to pass on their knowledge to me, I would not have understood the lo sbalzo *technique so easily."*

ALESSANDRO MARCHETTI

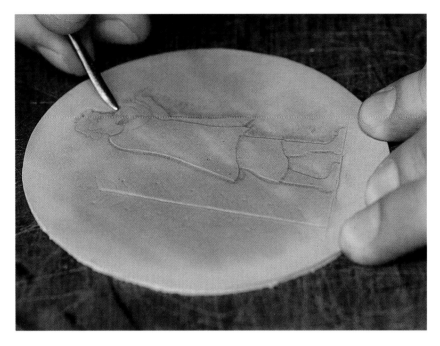

Left

Alessandro etches out a traced design onto a piece of undyed leather, later to be used in the center of one of Mara Santoni's bowls.

131

ANOTHER YOUNG CRAFTSPERSON in Tuscany who also learned leather work from an older generation of artisans is Alessandro Marchetti. His daily work used to be in a steel factory in Piombino making train wheels, but he often spent his evenings watching a leather tanner at work in Siena. For a year, the old man allowed him to watch, but not touch – then sent him away with a proclamation that he had now seen all there was to know. The skill he had learned is called *lo sbalzo* (thought to have originated in Spain) – the technique of pressing on leather to create a repeating decorative motif using very simple tools.

Like Patrizia, Alessandro began by making traditional belts, bags, and wallets. In the 1970s the development of machines to create the *sbalzo* effect on leather stimulated the "Tuscan leather boom," to such a degree that by 1992 there were only 10 people left in the whole of Italy still practicing the handmade technique. Alessandro was one of these. In addition to belts, bags, and wallets in the more traditional style, Alessandro has developed a unique application of his skill with his partner Mara (see pages 40–41). Her passion is for making pots, bowls, plates, and vases in the Etruscan *bucchero* tradition, and some of these are now decorated with leather. Both skills are ancient, handcrafted, and involve natural processes, and the subtle matte sheen common to both *bucchero* and leather make them ideal partners.

132

Above

Domenico is carefully
filling in the colors on one
of the side panels of a cart
that he is bringing back to
life. These panels show
mythical paladin scenes
which hark back to the
time of the crusades.

Opposite

Wheels and side panels
showing the polychromatic
painting that covers every
wooden surface of the
cart, including the under-
side.

DOMENICO DI MAURO & ANTONIO ZAPPALA, SICILIAN CARTS, SICILY

NORTH OF THE SICILIAN capital of Catania in Aci San'Antonio, another partnership of somewhat longer years is still thriving: Domenico di Mauro and Antonio Zappala, both approaching their 90s, are still handpainting carts (*carri*), an artisan craft steeped in traditional Sicilian folklore. Domenico has been painting them since he was 12 and began his own business in 1927 at the age of 14. Upright and sprightly – his moustache bristling, his eyes twinkling – he bears a striking resemblance to Einstein, whose picture is cheekily tacked on the wall of the workshop. Still enthusiastic and undoubtedly young at heart, he continues to strive, he says, to improve his technique and better his style. Renowned throughout Sicily, his brilliantly decorated carts have appeared in state and royal processions in Florence, Paris, London, Canada, Japan, and America and have won him prestigious awards, but he has never rested on his laurels.

Certainly, his is a dying art form. When he started out, there were 20 or so painters in the locality, but now he and his brother-in-law are the only ones left. Recently, however, a growing appreciation of tradition has resulted in a demand for renovating family "heirlooms," and they have consequently enjoyed

a boom in business. Domenico is delighted to have as apprentices two women who obviously share his passion for this aspect of their Sicilian heritage: Laura Paladino, who comes up from Catania a few days a week, and Maria Louisa, who studies his every brushstroke and absorbs his every word with the true dedication of a student for their master. Their eagerness and enthusiasm is reciprocated as there is no family successor to whom he can teach more than 70 years of knowledge and skill (his daughter is a lawyer). The tradition of decorated carts began at the end of the 18th century and has always been particularly popular in Sicily. At first, sacred images were painted on the panels "for protection" but gradually, as cart owners vied with each other, the variety of scenes depicted became more inventive. Bold primary colors, predominantly red, lend an overall gaiety and festive air, and not a centimeter of the surface is left undecorated. Wheels, handles, and framework are covered with a dense pattern of flowers and geometric designs, while the panels and base are the focus for the storytelling pictures. Carts are painted in one of two styles: *Rusticana* depicting 19th-century nostalgic pastoral scenes, or *Paladini* with illustrations of mythical Saracen battles and fairy tale legends of romance and chivalry.

Right

Cart side panel detail showing one of the 19th-century *Rusticana* or nostalgic pastoral scenes.

134

ANTONIO SAMARELLI, TERRACOTTA WHISTLES, PUGLIA

ANOTHER FOLKLORE TRADITION, but one that has moved with the times, is that of the *fischietto* or terracotta whistle. The small Puglian town of Rutigliano, just inland from Bari in the heel of Italy, has elevated the making of whistles to an art form, to the extent of founding a "whistle museum" and holding an annual competition which has been running since 1989.

Whistles and the act of whistling are deeply rooted in all aspects of primitive culture, including religion, superstition, and economic and social affairs. Nowadays in Rutigliano they are more associated with carnival and the feast day of the local saint, Sant' Antonio Abato. Made from terracotta, they range in design from the simple one-figure models of clowns, Pierrots, curvy ladies, clerical, and political figures to complex scenes of satirical, lyrical, or comical significance. The naïve modeling is an intrinsic part of their appeal and carnivalesque quality: the simpler whistles are reminiscent of British end-of-the-pier type ephemera, whereas the more complex models range from "Wallace and Gromit–type" characters and scenarios to quite beautiful fairy tale scenes and abstract contemporary art.

The annual competition in Rutigliano has inspired incredible creativity – many of the creations are instruments of social and political comment – and because it is national, artistic links with other parts of the country have developed, most notably with Bassano and Vicenza in the north.

Above

Antonio Samarelli, son of a local potter, has been making *fischietti* (whistles) since 1970. Here he is blowing a police-man whistle, one of many colorful folklore whistles that he makes in a gently satirical design.

136

ARTIGIANATI ARTESINI, DECORATED EGGS, TRENTINO ALTO-ADIGE

"Eggs are symbolic of fertility and rebirth and are presented in the church at Easter, as well as on other occasions such as birth, marriage, and death. I hope that we do not forget to teach our children about these symbols which link us all with our history and folklore traditions."

EDITH GRISENTI

GOOSE, TURKEY AND DUCK EGGS, painted with geometric or floral patterns, are a rich, colorful, and immensely skillful example of a folklore tradition. Decorated eggs are associated with the eastern European orthodox churches of Greece and Russia, which are more attuned to traditional pagan rituals of the populace, as symbols of fertility and rebirth. Bohemia is acknowledged as the origin of the most fabulously and ingeniously decorated eggs, but other countries such as Germany, Austria, Switzerland, and Italy, which have adopted the art form, have developed their own traditions. In the mountain area of Bolzano, formerly part of the Austro-Hungarian Empire, Austrian-born Edith Grisenti and her colleagues Lorenz Marmsaler and Elsa Kondrak, at the Artigianati Artesini have all evolved very individual styles: richly colored designs, creamy colored delicate patterns suggestive of lace, and embroidery, dainty miniatures of spring flowers and butterfly wings — all are applied freehand with the finest of brushes and the greatest of patience.

Edith is self-taught and has been developing her skills for more than 30 years. She says that goose eggs are the best to work with because they are the most robust and will even stand up to etching and engraving. Age-old techniques are used to ensure the very best results, such as cleaning the eggs with onions to keep them grease-free (and therefore receptive to paint), as well as more modern methods such as using clear nail varnish as a base coat on which to paint without the colors running. A final coat of wax protects the design and lends a natural matte sheen.

Opposite

A bowl of richly colored "batik" or "ethnic" designs created by Lorenz Marmsaler, one of the artisans who make up Edith Grisenti's little group of Artigianati Artesini.

Below

A deeply colored egg with delicate lacelike patterns decorated by Edith Grisenti.

137

OPEN ART, FRESCO ART, LOMBARDY

138

USING A MUCH BROADER brush stroke, Roberta Ricchini paints frescoes at her Open Art workshop in the industrial heart of Brescia. The art of painting on walls is one of the oldest in the world, dating back to prehistoric times. The most sustained use of fresco was in Italy between 1300 and 1800, hence the Italian terminology. Generations of gifted artists applied their artistry to walls and ceilings with Michelangelo's *Last Judgment* for the Sistine Chapel possibly the most famous fresco cycle ever painted. The durability of this art was achieved by painting directly on to wet plaster (*fresco* means "fresh"), the pigment of the paint chemically bonding with the surface. However, Roberta manages to practice fresco for different clients as far afield as Germany and France without leaving her workshop – by using a method called *fresco strappo* (literally, "freshly torn off"), which makes the fresco portable.

Roberta has no family artisan tradition, but does have a passion for her country's heritage as well as an obvious artistic talent. Having studied at art school in Milan, she worked as a professional accountant for five years before giving in to her artistic yearnings as a way of making a living. Her love of interior design and decoration combined with an admiration for the Italian art of fresco led her to set up this small business. She and her colleagues are able to copy almost any design, but clients generally commission traditional still-life subjects and pastoral scenes for their interiors. No work on canvas or paper has the same subtle, dusted appearance as an old fresco.

Above
(left to right)

Preparation of the colors used for painting the fresco; painting the "fine" version; the distressed stage.

Right

Two classical urns ready to be transferred onto muslin which will be pasted over the frescoes with a special glue before being peeled off to make the *fresco strappo* transfer.

map ~ where the artisans are located

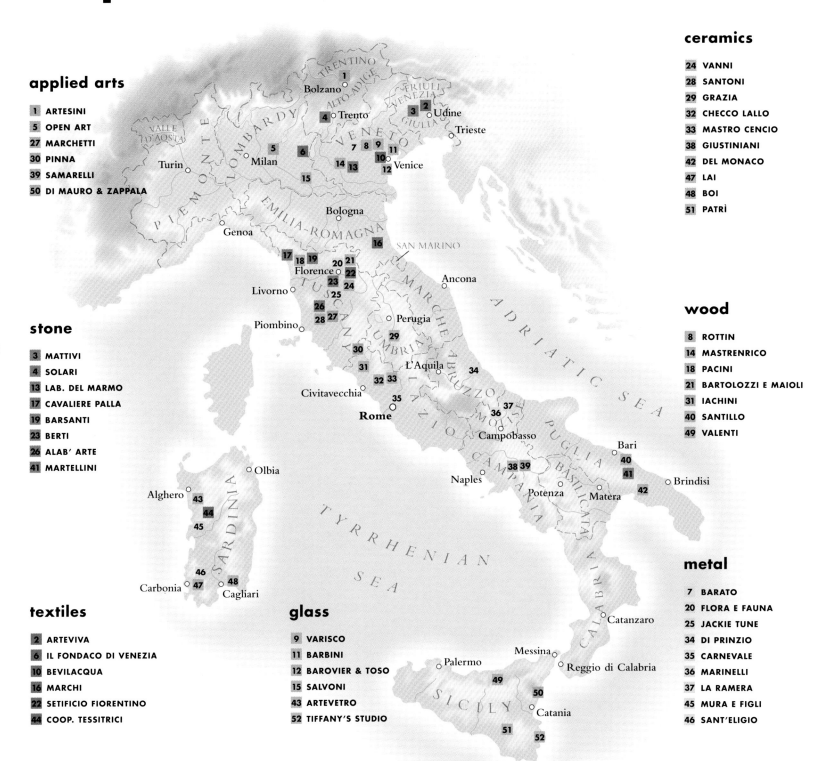

applied arts

1 ARTESINI
5 OPEN ART
27 MARCHETTI
30 PINNA
39 SAMARELLI
50 DI MAURO & ZAPPALA

stone

3 MATTIVI
4 SOLARI
13 LAB. DEL MARMO
17 CAVALIERE PALLA
19 BARSANTI
23 BERTI
26 ALAB' ARTE
41 MARTELLINI

textiles

2 ARTEVIVA
6 IL FONDACO DI VENEZIA
10 BEVILACQUA
16 MARCHI
22 SETIFICIO FIORENTINO
44 COOP. TESSITRICI

glass

9 VARISCO
11 BARBINI
12 BAROVIER & TOSO
15 SALVONI
43 ARTEVETRO
52 TIFFANY'S STUDIO

ceramics

24 VANNI
28 SANTONI
29 GRAZIA
32 CHECCO LALLO
33 MASTRO CENCIO
38 GIUSTINIANI
42 DEL MONACO
47 LAI
48 BOI
51 PATRÌ

wood

8 ROTTIN
14 MASTRENRICO
18 PACINI
21 BARTOLOZZI E MAIOLI
31 IACHINI
40 SANTILLO
49 VALENTI

metal

7 BARATO
20 FLORA E FAUNA
25 JACKIE TUNE
34 DI PRINZIO
35 CARNEVALE
36 MARINELLI
37 LA RAMERA
45 MURA E FIGLI
46 SANT'ELIGIO

the artisans

All artisans marked with an asterisk (★) prefer to be contacted by telephone in order to make an appointment for a visit. The remainder can be visited without prior arrangement. Opening hours vary considerably throughout the year so it is always prudent to check beforehand.

stone

ALAB'ARTE
Giorgio Finazzo and Roberto Chiti
Via Orti S Agostino 28
56048 Volterra (Pi)
Tuscany
Tel/Fax: (39) 0588 87968
Alabaster

ARTE BARSANTI
Via Val Di Lima 34
55021 Bagni di Lucca (Lu)
Tuscany
Tel: (39) 0583 87882
Fax: (39) 0583 808371
Plaster of Paris sacred statues

CLAUDIO BERTI
Via Isaac Newton 80
50018 Scandicci (Fi)
Tuscany
Tel/Fax: (39) 055 757388
Works in pietra dura

**LABORATORIO DEL MARMO/
GABRIELE BUBOLI**
Viale Spalato 38 a
Montagnana (Pd)
Veneto
Tel (39) 0429 804006
Fax: (39) 0429 82919
Email: ldm@netbusiness.it
www.labmarmo.com
Stonework

MATTIVI MARMI
Via Solteri 5
38100 Trento
Trentino Alto Adige
Tel: (39) 0461 821531
Fax: (39) 0461 429525
Stonework

GIOVANNI MARTELLINI
Martina Franca (Ta)
Puglia
Tel: (39) 080 4831432★
Stonework

CAVALIERE FERDINANDO PALLA
P Carducci Giosue' 18
55045 Pietrasanta (Lu)
Tuscany
Tel/Fax: (39) 0584 70224
Marble and pietra dura

VALTER SOLARI
Dignano (Ud)
Friuli Venezia-Giulia
Tel: (39) 0432 951529 ★
Mosaics

ceramics

MASSIMO BOI
Quartu S'Elena (Ca)
Sardinia
Tel: (39) 070 807456 ★
Modern ceramics

CHECCO LALLO
Via dei Pilari
Vetralla (Vt)
Lazio
Rustic pottery

DEL MONACO
Via S Sofia 2/4
74023 Grottaglie (Ta)
Puglia
Tel: (39) 099 5661023
Fax: (39) 099 5667522
Traditional Puglian ceramics

BOTTEGA NICOLA GIUSTINIANI
Elvio Sagnella
Via S Donato 10
82030 San Lorenzello (Bn)
Campania
Tel/Fax: (39) 0824 861700
18th-century ceramics

UBALDO GRAZIA
Via Tiburtina 181
06053 Deruta (Pg)
Umbria
Tel: (39) 075 9710201
Fax: (39) 075 972018
Email: ugrazia@tin.it
www.ubaldograzia.com
Ceramics

ALESSANDRO LAI
Iglésias (Ca)
Sardinia
Tel/Fax: (39) 0781 33850 ★
Archaic pottery

STUDIO D'ARTE MASTRO CENCIO
Via SS Giovanni e Marciano
Martiri 14
01033 Civita Castellana (Vt)
Lazio
Ceramics

MAURIZIO PATRÌ
Via Roma 17
95041 Caltagirone (Ct)
Sicily
Tel: (39) 0933 26850
Traditional Sicilian ceramics

MARA SANTONI
Via Moncini 40
58024 Massa Marittima (Gr)
Tuscany
Tel/Fax: (39) 0566 901551
Bucchero

LUCA VANNI
Via Europa 7
50023 Impruneta (Fi)
Tuscany
Tel: (39) 055 2312247
Terracotta

wood

BARTOLOZZI E MAIOLI
Via Maggio 13r/Via Toscanella 8
Florence (Fi)
Tuscany
Tel: (39) 055 282675
Fax: (39) 055 217870
Email: bem@worldlink.it
Woodcarving and restoration

ANTONIO IACHINI
Via Torre di Lavello 14
01017 Tuscania (Vt)
Lazio
Tel: (39) 0761 434033
Marquetry and restoration

LORENZO PACINI
Via Sottomonte 27/a
55060 Guamo (Lu)
Tuscany
Tel: (39) 0583 403354
Fax: (39) 0583 403382
Reproduction furniture

MASTRENRICO
Via Borgo Eniano 109
Montagnana (Pd)
Veneto
Tel: (39) 0429 82744
Utility ware

VALTER ROTTIN
Piazzale Duca d'Aosta 28
31100 Treviso (Tv)
Veneto
Tel: (39) 0422 542266
Decorative pieces and furniture

OSVALDO SANTILLO
Pietralcina
Benevento
Campania
Tel: (39) 0824 991972 ★
Baskets

141

PINO VALENTI
Collesano
Palermo
Sicily
Tel: (39) 0921 661895 ★
Intarsia and mosaics

metal

ARTIGIANI SANT'ELIGIO
Via Cattaneo
09016 Iglésias (Ca)
Sardinia
C/o Tel/Fax: (39) 0781 33850
Silver-plated items

FLORA E FAUNA
Sdrucciolo Pitti 22
50125 Florence (Fi)
Tuscany
Tel: (39) 055 2382162
Fax: (39) 055 2382229
Email: floraefauna@libero.it
Decorative brass

FONDERIA CARNEVALE
Via dei Sabelli 122
00185 Roma
Tel: (39) 06 4453363
Fax: (39) 06 4453605
Ornamental cast iron

FONDERIA MARINELLI
Via F D'Onofrio 14
86081 Agnone (Is)
Molise
Tel/Fax: (39) 0865 78235
Bells

FRATELLI BARATO, ARTIGIANI DEL RAME
Via Vanzo Nuovo 18
Camisano Vicentino (Vi)
Veneto
Tel: (39) 0444 410796
Copperwork

LA RAMERA
Corso V Emanuele 252
86081 Agnone (Is)
Molise
Tel: (39) 0865 779086
Fax: (39) 0865 779853
Copper utensils

VITTORIO MURA E FIGLI
Viale Azuni 4
09075 Santu Lussúrgiu (Or)
Sardinia
Tel: (39) 0783 550726
Metalwork and knives

RAFAELE DI PRINZIO
Via S Eufemia 103
66010 Fara Filiorum Petri (Ch)
Abruzzo
Tel: (39) 0871 70172
Wrought-iron metalwork

JACKIE TUNE
Casole d'Elsa
Siena
Tuscany
Tel: (39) 0577 960298 ★
Decorative metal

textiles

ANTICA STAMPERIA MARCHI
Via C Battisti 15
47038 Santarcangelo (Ra)
Emilia Romagna
Tel: (39) 0541 626018
Hand-printed linens

ANTICO SETIFICIO FIORENTINO
Via Bartolini 4
50124 Florence (Fi)
Tuscany
Tel: (39) 055 213861
Fax: (39) 055 218174
Luxury silk

ARTEVIVA
Via G Muratti 27
33100 Udine (Ud)
Friuli
Tel: (39) 0432 510529
Woven textiles and tapestry

COOP. TESSITRICI
Artigianato di Villanova
Via Nazionale 71
07019 Villanova Monteleone (Sa)
Sardinia
Tel: (39) 079 960474
Traditional woven items

IL FONDACO DI VENEZIA
Via Zanardelli 58
25083 Gardone Riviera (Bs)
Lombardia
Tel: (39) 0365 22040
Fax: (39) 0365 21214
Email: fondaco@mail.gsnet.it
Hand-printed velvet

TESSITURA BEVILACQUA
Santa Croce 1320
30135 Venice
Tel: (39) 041 721566
Fax: (39) 041 5242302
Email: bevilacqua@luigi-bevilacqua.com
www.luigi-bevilacqua.com
Silk and velvet

glass

ARTEVETRO
Via Palomba 65
07041 Alghero (Sa)
Sardinia
Tel: (39) 079 981608
Fused and leaded glass

FRATELLI BARBINI
Calle Bertolini 36
Murano
30141 Venice
Tel/Fax: (39) 041 739777
Venetian mirrors

BAROVIER & TOSO
Fonderia Vetrai 28
Murano
30141 Venice
Tel: (39) 041 739049
Fax: (39) 041 5274385
Email: barovier@barovier.com
Classic glass

PIER LORENZO SALVONI
Pompiano (Bs)
Lombardia
Tel: (39) 030 946 1240 ★
Decorative glass

TIFFANY'S STUDIO AND ART WORK
Via Maestranza 76
96100 Siracusa (Ct)
Sicily
Tel: (39) 0931 463649
Tiffany-style and fused glass

BOTTEGA VARISCO
Via Nervesa della Battaglia 59
31100 Treviso (Tv)
Veneto
Tel: (39) 0422 300980
Fax: (39) 0422 306611
Email: cristallivarisco@libero.it
www.cristallivarisco.it
Engraved glass

applied arts

ARTIGIANATI ARTESINI
Bolzano
Trentino Alto-Adige
Tel: (39) 0471 978590★
Decorated eggs

DOMENICO DI MAURO AND ANTONIO ZAPPALA
Aci Sant'Antonio
Catania
Sicily
Tel: (39) 095 7921768 ★
Sicilian carts

OPEN ART
Via Rose di Sotto 7
25126 Brescia (Bs)
Lombardia
Tel/Fax: (39) 030 2906300
Fresco art

ALESSANDRO MARCHETTI
Via Moncini 40
58024 Massa Marittima (Gr)
Tuscany
Tel/Fax: (39) 0566 901551
Leather work

PATRIZIA PINNA
Via G Selvi 7
58010 Sorano (Gr)
Tuscany
Tel: (39) 0564 633647
Leather work

ANTONIO SAMARELLI
Via Noicattaro 233/245
Rutigliano (Ba)
Puglia
Tel: (39) 080 4761437
Terracotta whistles

142

index

143

acknowledgments

I would like to thank first and foremost **all of the artisans** that I visited during the preparation of this book. Their kindness and frequent hospitality was especially appreciated since they were often bemused but pleasantly surprised by my interest in them and their work. In every way, this book is dedicated to them and to the countless others who carry on working diligently and modestly in every corner of Italy, whether in the big cities or deep in the countryside. I tried to ensure that as many regions as possible were represented, but apologies to those that are not.

Secondly, I must thank **Edoardo Betti** and **Alessandra Smith** from the **Italian Tourist Board** in London, whose help made the project much less difficult than it might have been. Their assistance and that of the many tourist boards in Italy will, hopefully, reap rewards by helping to create a better understanding of the extremely valuable contribution that all of Italy's artisans make to their country's culture and, indeed, tourism potential.

I must also thank **Patrizia Enne** in Sardinia for her assistance in getting to know the artisans in Iglésias, **Mary Jane Cryan** (Vetralla) for the use of her wonderful library and my dear friends, **the Caponetti family,** who always welcome me back to their special home in Tuscania. Last but certainly not least, a big hug to **Debra Boraston,** my dear wife, who accompanied me and assisted in the writing of this book.